Lecture Notes in Energy 17

For further volumes:
http://www.springer.com/series/8874

Martin Junginger • Chun Sheng Goh
André Faaij
Editors

International Bioenergy Trade

History, status & outlook on securing sustainable bioenergy supply, demand and markets

Editors
Martin Junginger
Copernicus Instituut of Sustainable Development
Utrecht University
Utrecht, Netherlands

Chun Sheng Goh
Copernicus Instituut
Utrecht University
Utrecht, Netherlands

André Faaij
Copernicus Instituut
Utrecht University
Utrecht, Netherlands

ISSN 2195-1284 ISSN 2195-1292 (electronic)
ISBN 978-94-007-6981-6 ISBN 978-94-007-6982-3 (eBook)
DOI 10.1007/978-94-007-6982-3
Springer Dordrecht Heidelberg New York London

Library of Congress Control Number: 2013951141

Printed on acid-free paper

Springer is part of Springer Science+Business Media (www.springer.com)

Contents

Chapter 1
A General Introduction to International Bioenergy Trade

André Faaij, Martin Junginger, and Chun Sheng Goh

Abstract The development of functional international markets for bioenergy has become an essential driver to develop bioenergy potentials, which are currently under-utilised in many regions of the world. Technical potential of bioenergy may be as large as 500 EJ/yr by 2050. However, large uncertainty exists about important factors such as market and policy conditions that affect this potential. Potential deployment levels by 2050 could lay in the range of 100–300 EJ/yr. Realizing this potential represents a major challenge but would substantially contribute to the world's primary energy demand in 2050. The possibilities to export biomass-derived commodities for the world's energy market can create important socioeconomic development incentives for rural communities. But bioenergy markets are still immature, relying on policy objectives and incentives, that prove to be erratic in many cases. Further improvement is needed to develop both supply and demand in a balanced way and avoid distortions and instability that can threaten investments. Furthermore, it is necessary to develop and exploit biomass resources in a sustainable way and to understand what this means in different settings. In some markets, prices of biomass resources are volatile, including indirect effects on price of raw material prices for e.g. the forest industry as well as on food. Sustainability demands serve as a starting point for policies supporting bioenergy in many countries. The proliferation of initiatives registered worldwide to develop and implement sustainability frameworks and certification systems for bioenergy, can lead to a fragmentation of efforts. This asks for harmonization and for international collaboration.

A. Faaij (✉) • M. Junginger • C.S. Goh
Copernicus Institute, Utrecht University, Utrecht, The Netherlands
e-mail: a.p.c.faaij@uu.nl; h.m.junginger@uu.nl; c.s.goh@uu.nl

M. Junginger et al. (eds.), *International Bioenergy Trade: History, status & outlook on securing sustainable bioenergy supply, demand and markets*, Lecture Notes in Energy 17,
DOI 10.1007/978-94-007-6982-3_1, © Springer Science+Business Media Dordrecht 2014

1.1 Background of the Book

Biomass is the most important renewable energy source today and expected to play a major role in medium to longer term in replacing fossil fuels and reducing greenhouse gas emissions. Up to 2008, ambitions for biomass use for energy were high in many countries, for the EU and also on a global basis, give a variety of policy objectives and long term energy scenario's (Hunt et al. 2007). A reliable supply and demand of bioenergy is vital to develop stable market activities. As a result of various targets and incentives, the trade of global bioenergy commodities, such as ethanol, biodiesel and wood pellets has been growing exponentially in the past decade, and have by 2011 reached true "commodity" volumes, i.e. tens of millions of tonnes traded each year, and billions (both in US$/€) of annual turnover.

The development of functional international markets for bioenergy has become an essential driver to develop bioenergy potentials, which are currently underutilised in many regions of the world. This is true for both residues and for dedicated biomass production (through energy crops or multifunctional systems such as agroforestry). The possibilities to export biomass-derived commodities for the world's energy market can provide a stable and reliable demand for rural communities, thus creating important socioeconomic development incentives and market access. International trade of biomass and biofuels and market development has been a major trend over the past years and a major stabilizing factor for the bioenergy sector worldwide (Faaij and Domac 2006).

The need to reduce GHG emissions, secure energy supplies and achieve rural development come together in bioenergy. Many national policies and global scenario's target bioenergy (for heat, power, fuels, as well as biomaterials) to make a major contribution to achieving such goals. The IPCC (Metz et al. 2007; Chum et al. 2011) and Global Energy Assessment (GEA 2012) highlight biomass use as a key mitigation option in about any relevant sector, highlighting that the energy potential may ultimately be developed to 200–400 EJ (compared to some 500 EJ global primary energy use today and an expected 1,000 EJ in the timeframe between 2050 and 2100). Biomass shares in total energy supply are hoped to achieve 20–30 % looking at long term strategies and scenarios of countries as well as on a global scale. Recent scenario's stress the importance of combining large scale bioenergy use with carbon capture and storage, because this is one of the few options available to achieve net negative emissions and keep a global 2 °C temperature change target in sight during this century; a target widely seen as essential to limit the damage of climate change (GEA 2012).

Given the expectations for a high bioenergy demand on a global scale, the pressure on available biomass resources will increase. Without the further development and mobilisation of biomass resources (e.g. through energy crops and better use of agro-forestry residues) and a well-functioning biomass market to assure a reliable and lasting supply, those ambitions may not be met. A lack of availability of good quality (and competitive) biomass resources has proven to be a structural showstopper for many market initiatives in the past. Then, in 2008, the whole bioenergy and biofuels sector has ended up in turmoil during due to a spike in food prices globally. Biofuels were perceived to cause major conflicts with food supplies and be a main driver for food price increases. In addition, the problem of indirect land use change (displaced food crops by biofuels will be produced on other land, leading ultimately to conversion of forest and nature areas and thus carbon losses and indirect GHG emissions), as well as other external damages of (possible) conflicts with biodiversity, water use and socio-economic development.

Partly as a result from this crisis, securing sustainable production of biomass and biofuels has become a key topic globally and is likely to remain a leading issue for years to come. Given that in various ways, new policies are now implemented to introduce sustainability criteria for bioenergy (e.g. the Renewable Energy Directive of the European Commission, various roundtables (sugar, soy, palm oil and on sustainable biofuels), market driven frameworks and various national proposals (Dam et al. 2010)), it is of vital importance to improve our understanding how the current proposals and their expected development will affect trade and markets of biomass and biofuels.

Furthermore, there is strong policy pressure to introduce and implement second generation biofuels on a large scale on shortest possible term to move away from using food crops as feedstock for biofuels. Task 40 acknowledged that a breakthrough and strong support for 2nd generation biofuels will have major impacts on markets and trade. The same may be true for new biochemicals by development of advanced biorefineries. Supplying large-scale second-generation plants with sufficient biomass will in most cases require advanced logistics and international sourcing of (lignocellulosic) feedstock. It is also observed that more and more investment is taking place in developing regions and economies in transition for biomass and biofuel production with export as a main target.

At the same time, the financial crisis strongly affects the investments in alternative energy in general. Whether the government support for renewable energy and bioenergy in particular will be sufficient to drive developments for bioenergy further in the coming years remains a major uncertainty at present.

On the positive side, various countries now have considerable experience with building biomass markets and developing available resources in balance with market demand. Over the past decade or so, international trade of biomass resources and biofuels has become indispensable in the portfolio of market parties and volumes traded worldwide increase at a very rapid pace with an estimated doubling of volumes in several markets over the past few years. In 2007, it was estimated that already 10 % of global bioenergy production involved internationally trade biomass resources and fuels. In specific markets (e.g. wood pellets) up to one third of the global demand is nowadays traded internationally. Many developing countries have a large technical agro-forestry residue potential as well as for dedicated energy plantations e.g. ethanol from sugar cane, pellets or wood plantations or agro-forestry systems (Smeets et al. 2007). Given the lower costs for land and labour in many developing countries, biomass production costs can be low, and thus offer an opportunity to export biomass to developed countries. The fact that those international markets are growing means that more and more resources are becoming available from regions were biomass use was low or absent so far and supply risks for biomass users have reduced due to more diverse and reliable supplies. International bioenergy trade is in that sense a stabilizing and enabling factor for developing the bioenergy option at large on a global scale.

In this context, an international working group was established in 2004 under the IEA Bioenergy Implementing Agreement: Task 40 on Sustainable International Bioenergy Trade. For market parties such as utilities, companies providing transport fuels, as well as parties involved in biomass production and supply (such as forestry companies), good understanding, clear criteria and identification of promising possibilities and areas are of key interest. Investments in infrastructure and conversion capacity rely on minimization of risks of supply disruptions (in terms of volume, quality as well as price).

By 2013, the Task 40 network has started its' 4th triennium. For the past 10 years, Task 40 has monitored the developments in international bioenergy trade, including the organization of a few dozen international workshops on biotrade-related topics, and the publication of over 100 studies, country reports, newsletters, etc. The amount of material produced over the years and insights gained in how biomass markets and international trade of biomass and biofuels has developed is therefore substantial. Besides that the group has produced overviews and analyses, also a large amount of practical experiences has been brought together in what works and what doesn't. Last but not least, based on all this, the work of the network delivered clear(er) views on how to proceed to build working sustainable international biomass markets in the future. Therefore, those lessons and insights are compiled into an easily accessible book publication.

This book is the first to cover basically all biomass commodities that are traded in any significant volumes in detail. The author team is compiled from industry & academia and from the main biomass and biofuel exporting & importing countries. Task 40 members include e.g. Canada and Brazil, two major exporters of wood pellets and ethanol, and the USA, Japan and 10 EU of which many actively trade biodiesel, ethanol and wood pellets. While naturally also trade outside the Task 40 member countries will be highlighted, our member countries encompass probably more than 90 % of energy-related trade of wood pellets, biodiesel and ethanol. Also, Task 40 members encompass a biomass producer and several biomass traders and large-scale end-users. Next to the industrial representatives, key academic institutes from amongst others Utrecht University, Imperial College London, Vienna University of Technology, Lappeenranta University, University of Campinas, Norwegian University of Life Sciences and many other research institutes on four continents.

This book is relevant for industry and policy makers as well as for NGO's and academia. It speaks for itself that growing bioenergy trade is relevant for market parties, but also policy makers will increasingly be interested in the growing global bioenergy markets, as increasingly, bioenergy/renewable energy targets will only be achievable with increasing trade volumes.

1.2 Global Sketch of the Current Role of Bioenergy in the Global Energy Supply and the Role of International Bioenergy Trade

Biomass provided about 10.2 % (50.3 EJ/year) of the annual global primary energy supply in 2008, from a wide variety of biomass sources feeding numerous sectors of society (see Table 1.1; Chum et al. 2011). The biomass feedstocks used are for more than 80 % are derived from wood (trees, branches, residues) and shrubs. The remaining bioenergy feedstocks came from the agricultural sector (energy crops, residues and by-products) and from various commercial and post-consumer waste and by-product streams (biomass product recycling and processing or the organic biogenic fraction of municipal solid waste (MSW)).

Global bioenergy use has steadily grown worldwide in absolute terms in the last 40 years, with large differences among countries. In 2006, China led all countries and used 9 EJ of biomass for energy, followed by India (6 EJ), the USA (2.3 EJ) and Brazil (2 EJ) (GBEP 2008). Bioenergy provides a relatively small but growing share of TPES (1–4 % in 2006) in the largest industrialized countries (grouped as the G8

Table 1.1 Examples of traditional biomass and modern bioenergy flows in 2008 according to the IEA (2010a) and Chum et al. (2011)

Type	Primary energy (EJ/year)	Approximate average efficiency (%)	Secondary energy carrier (EJ/year)
Traditional biomass used for bioenergy			
Accounted for in energy balance statistics (IEA 2010a, b)	30.7	10–20	3–6
Estimated for informal sectors (e.g., charcoal)	6–12		0.6–2.4
Total traditional biomass used for energy	*37–43*		*3.6–8.4*
Modern bioenergy (IEA 2010a)			
Electricity and CHP from biomass, MSW and biogas	4.0	32	1.3
Heat in residential and public/commercial buildings from solid biomass and biogas	4.2	80	3.4
Road transport fuels (ethanol and biodiesel)	3.1	65	1.9
Total modern bioenergy	*11.3*	*59*	*6.6*

Notes: The global primary biomass supply of 50.3 EJ for energy is composed primarily of solid biomass (46.9 EJ), biogenic MSW used for heat and CHP (0.58 EJ); biogas (secondary energy) for electricity and CHP (0.41 EJ) and for heating (0.33 EJ). Delivered (secondary energy) ethanol, biodiesel, and other transport fuels (e.g., ethers) made up 1.9 EJ. Examples of specific flows: output electricity from biomass was 0.82 EJ (biomass power plants including pulp and paper industry surplus, biogas and MSW) and output heating from CHP was 0.44 EJ. Modern residential heat consumption was calculated by subtracting the IEA estimate of traditional use of biomass (30.7 EJ) from the total residential heat consumption (33.7 EJ)

countries: the USA, Canada, Germany, France, Japan, Italy, the UK and Russia). The use of solid biomass for electricity production is particularly important in pulp and paper plants and in sugar mills. Bioenergy's share in total energy consumption is generally increasing in the G8 countries through the use of modern biomass forms (e.g., co-combustion or co-firing for electricity generation, space heating with pellets) especially in Germany, Italy and the UK.

By contrast, in 2006, bioenergy provided 5–27 % of TPES in the largest developing countries (China, India, Mexico, Brazil and South Africa), mainly through the use of traditional forms, and more than 80 % of TPES in the poorest countries. The bioenergy share in India, China and Mexico is decreasing, mostly as traditional biomass is substituted by kerosene and liquefied petroleum gas within large cities. However, consumption in absolute terms continues to grow. This trend is also true for most African countries, where demand has been driven by a steady increase in wood fuels, particularly in the use of charcoal in booming urban areas (GBEP 2008).

Global trade in biomass feedstocks (e.g., wood chips, raw vegetable oils, agricultural residues) and especially of energy carriers from modern bioenergy (e.g., ethanol, biodiesel, wood pellets) is growing rapidly. While practically no liquid biofuels or wood pellets were traded in 2000, the world net trade of liquid biofuels amounted to 120–130 PJ in 2009, compared to about 75 PJ for wood pellets. Larger quantities of these products are expected to be traded internationally in the future, with Latin America and sub-Saharan Africa as potential net exporters and North America, Europe and Asia expected as net importers (Heinimö and Junginger 2009). Trade can therefore become an important component of the sustained growth of the bioenergy sector. In 2008, around 9 % of global biofuel production was traded internationally. Production and trade of these three commodities are discussed in more detail below.

Global fuel *ethanol production* grew from around 0.375 EJ in 2000 to more than 1.6 EJ in 2009 (Lamers et al. 2011). The USA and Brazil, the two leading ethanol producers and consumers, accounted for about 85 % of the world's production. In the EU, total consumption of ethanol for transport in 2009 was 94 PJ (3.6 Mt), with the largest users being France, Germany, Sweden and Spain (Lamers et al. 2011). Data related to fuel *bioethanol trade* are imprecise on account of the various potential end uses of ethanol (i.e., fuel, industrial and beverage use) and also because of the lack of proper codes for biofuels in global trade statistics. As an estimate, a net amount of 40–51 PJ of fuel ethanol was traded in 2009 (Lamers et al. 2011).

World *biodiesel production* started below 20 PJ in 2000 and reached about 565 PJ in 2009 (Lamers et al. 2011). The EU produced 334 PJ (roughly two-thirds of the global production), with Germany, France, Spain and Italy being the top EU producers. EU27 biodiesel production rates levelled off towards 2008. The intra-European biodiesel market has become more competitive, and the 2009 overcapacity has already led to the closure of (smaller, less vertically integrated, less efficient, remote, etc.) biodiesel plants in Germany, Austria and the UK. As shown in Fig. 2.9, other main biodiesel producers include the USA, Argentina and Brazil. Biodiesel consumption in the EU amounted to about 403 PJ (8.5 Mt), with Germany and France consuming almost half of this amount. Net international *biodiesel trade* was below 1 PJ before 2005 but grew very fast from this small base to more than 80 PJ in 2009, as shown in Fig. 2.9 (Lamers et al. 2011).

Production, consumption and trade of *wood pellets* have grown strongly within the last decade and are comparable to ethanol and biodiesel in terms of global trade volumes. As a rough estimate, in 2009, more than 14 Mt of wood pellets were produced (Lamers et al. 2012) primarily in 30 European countries, the USA and Canada. Consumption was high in many EU countries and the USA. The largest EU consumers were Sweden (1.8 Mt), Denmark, the Netherlands, Belgium, Germany and Italy (roughly 1 Mt each). Main *wood pellet trade* routes lead from Canada and the USA to Europe (especially Sweden, the Netherlands and Belgium) and to the USA. In 2009, other minor trade flows were also reported, for example, from Australia, Argentina and South Africa to the EU. Canadian producers also started to export small quantities to Japan. Total imports of wood pellets by European countries in 2009 were estimated to be about 3.9 Mt, of which about half can be assumed to be intra-EU trade (Sikkema et al. 2011).

1.3 Main Drivers and Major Barriers for Future Development in International Bioenergy Markets and Trade

Support policies have strongly contributed in past decades to the growth of bioenergy for electricity, heat and transport fuels. However, several reports also point out the costs and risks associated with support policies for biofuels. According to the IEA World Energy Outlook (IEA 2010b), the annual global government support for biofuels in 2009, 2008 and 2007 was USD_{2009} 20 billion, 17.5 billion and 14 billion, respectively, with corresponding EU spending of USD_{2009} 7.9 billion, 8.0 billion and 6.3 billion and corresponding US spending of USD_{2009} 8.1 billion, 6.6 billion and 4.9 billion. The US spending was driven by energy security and fossil fuel import reduction goals. Concerns about food prices, GHG emissions and environmental impacts have also led to many countries rethinking biofuels blending targets. For example, Germany revised its blending target for 2009 downward from 6.25 to 5.25 %. Addressing these concerns led also to the incorporation of environmental and social sustainability criteria for biofuels in the EU Renewable Energy Directive. Although seemingly effective in supporting domestic farmers, the effectiveness of biofuel policies in reaching the climate change and secure energy supply objectives is coming under increasing scrutiny. It has been argued that these policies have been costly and have tended to introduce new distortions to already severely distorted and protected agricultural markets, at both domestic and global levels. This has not tended to favour an efficient international production pattern for biofuels and their feedstocks (FAO 2008). An overall biomass strategy would have to consider all types of use of food and non-food biomass.

The main drivers behind government support for the sector have been concerns over climate change and energy security as well as the desire to support the agricultural sector through increased demand for agricultural products (FAO 2008). According to the REN21 a total of 69 countries had one or several biomass support policies in place in 2009 (REN21 2010).

Governments are stressing the importance of ensuring sufficient climate change mitigation and avoiding unacceptable negative effects of bioenergy as they implement regulating instruments. For example, the Renewable Energy Directive (European Commission 2009) provides mandatory sustainability requirements for liquid transport fuels. Also, in the USA, the Renewable Fuel Standard—included in the 2007 Energy Independence and Security Act mandates minimum GHG reductions from renewable fuels, discourages use of food and fodder crops as feedstocks, permits use of cultivated land and estimates (indirect) LUC effects to set thresholds of GHG emission reductions for categories of fuels. The California Low Carbon Fuel Standard set an absolute carbon intensity reduction standard and periodic evaluation of new information, for instance, on indirect land use impacts. Other examples are the UK Renewable Transport Fuel Obligation, the German Biofuel Sustainability Ordinance, and the Cramer Report (The Netherlands). With the exception of Belgium, no mandatory sustainability criteria for solid biomass (e.g., wood pellets) have been implemented—the European Commission will review this at the end of 2011.

The development of impact assessment frameworks and sustainability criteria involves significant challenges in relation to methodology, process development and harmonization. As of a 2010 review, nearly 70 ongoing certification initiatives exist to safeguard the sustainability of agriculture and forestry products, including those used as feedstock for the production of bioenergy (van Dam et al. 2010). Within the EU, a number of initiatives started or have already set up certification schemes in order to guarantee a more sustainable cultivation of energy crops and production of energy carriers from modern biomass. Many initiatives focus on the sustainability of liquid biofuels including primarily environmental principles, although some of them, such as the Council for Sustainable Biomass Production and the Better Sugarcane Initiative, the Roundtable for Sustainable Biofuels (RSB) and the Roundtable for Responsible Soy, include explicit socioeconomic impacts of bioenergy production. Principles such as those from the RSB have already led to a Biofuels Sustainability Scorecard used by the Inter-American Development Bank for the development of projects.

The proliferation of standards that has taken place over the past 4 years, and continues, shows that certification has the potential to influence local impacts related to the environmental and social effects of direct bioenergy production. Many of the bodies involved conclude that for an efficient certification system there is a need for further harmonization, availability of reliable data, and linking indicators at micro-, meso- and macro- levels (van Dam et al. 2010). Considering the multiple spatial scales, certification should be combined with additional measurements and tools at regional, national and international levels.

The role of bioenergy production in iLUC is still uncertain (Chum et al. 2011; Wicke et al. 2012) and the use of iLUC factors obtained by model analyses in legislation is so far not implemented; current initiatives have rarely captured impacts from iLUC in their standards, and the time scale becomes another important variable in assessing such changes. Although recent discussions focus more on good practices to prevent iLUC risks, this remains a contentious issue and major point of attention for the coming years. The same is true for management of forest resources and avoiding unsustainable use leading to high "carbon debts", although also in that discussion field good management practices are essential to achieve good (environmental) performance of biomass sourcing and use (Jonker et al. 2013; Lamers et al. 2013). Addressing unwanted LUC requires overall sustainable agricultural production and good governance first of all, regardless of the end use of the product or of the feedstocks.

The review of developments in biomass use, markets and policy shows that bioenergy has seen rapid developments over the past years. The use of modern biomass for liquid and gaseous energy carriers is growing, in particular biofuels (with a 37 % increase from 2006 to 2009). Projections from the IEA, among others, but also many national targets, count on biomass delivering a substantial increase in the share of RE. Nevertheless, many barriers remain to developing well-working commodity trading of biomass and biofuels that at the same time meets sustainability criteria.

The policy context for bioenergy, and in particular biofuels, in many countries has changed rapidly and dramatically in recent years. The debate on food versus fuel competition and the growing concerns about other conflicts have resulted in a strong push for the development and implementation of sustainability criteria and frameworks as well as changes in temporization of targets for bioenergy and biofuels. Furthermore, the support for advanced biorefinery and second-generation biofuel options shifts attention in the market with expectedly profound implications for biomass demand and market players involved.

Persistent policy and stable policy support has been a key factor in building biomass production capacity and working markets, required infrastructure and conversion capacity that gets more competitive over time. These conditions have led to the success of the Brazilian programme to the point that ethanol production costs are lower than those of gasoline. Brazil achieved an energy portfolio mix that is substantially renewable and that minimized foreign oil imports. Sweden, Finland, and Denmark also have shown significant growth in renewable electricity and in management of integrated resources, which steadily resulted in innovations such as industrial symbiosis of collocated industries. The USA has been able to quickly ramp up production with the alignment of national and sub-national policies for power in the 1980s and for biofuels in the 1990s to present, as petroleum prices and instability in key producing countries increased; however, as oil prices decreased, policy support and bioenergy production decreased for biopower and is increasing again with environmental policies and sub-national targets.

Countries differ in their priorities, approaches, technology choices and support schemes for further development of bioenergy. Although this means increased complexity of the bioenergy market, this also reflects the many aspects that affect bioenergy deployment—agriculture and land use, energy policy and security, rural development and environmental policies. Priorities, stage of development and geographic access to the resources, and their availability and costs differ widely from country to country.

As policies surrounding bioenergy and biofuels become more holistic, using sustainability demands as a starting point is becoming an overall trend. This is true for the EU, the USA and China, but also for many developing countries such as Mozambique and Tanzania. This is a positive development but is by no means settled. The 70 initiatives registered worldwide by 2009 to develop and implement sustainability frameworks and certification systems for bioenergy and biofuels, as well as agriculture and forestry, can lead to a fragmentation of efforts (van Dam et al. 2010). The needs for harmonization and for international and multilateral collaboration and dialogue are widely stressed at present.

1.4 Possible Future Potentials and Future Deployment for Bioenergy and Implications for International Markets

The IPCC concluded in its Special Report on Renewable Energy the following on the possible future role of Bioenergy (Chum et al. 2011): biomass is likely to remain the largest RE sources for the first half of this century. There is considerable growth potential, but developing that potential in a sustainable way comes with a large number of preconditions.

- Assessments in the recent literature show that the technical potential of biomass for energy may be as large as 500 EJ/year by 2050. However, large uncertainty exists about important factors such as market and policy conditions that affect this potential.
- Potential deployment levels by 2050 could lay in the range of 100–300 EJ/year. Realizing this potential represents a major challenge but would make a substantial contribution to the world's primary energy demand in 2050.
- Bioenergy has significant potential to mitigate GHGs if resources are sustainably developed and efficient technologies are applied. Certain current systems and key future options including perennial crops, forest products and biomass residues and wastes, and advanced conversion technologies, can deliver significant GHG mitigation performance—an 80–90 % reduction compared to the fossil energy baseline. However, land conversion and forest management that lead to a large loss of carbon stocks and iLUC effects can lessen, and in some cases more than neutralize, the net positive GHG mitigation impacts.
- In order to achieve the high biomass potential deployment levels, increases in competing food and fibre demand must be moderate, land must be properly managed and agricultural and forestry yields must increase substantially. Expansion of bioenergy in the absence of monitoring and good governance of land use carries the risk of significant conflicts with respect to food supplies, water resources and biodiversity, as well as a risk of low GHG benefits. Conversely, implementa-

tion that follows effective sustainability frameworks could mitigate such conflicts and allow realization of positive outcomes, for example, in rural development, land amelioration and climate change mitigation, including opportunities to combine adaptation measures.
- The impacts and performance of biomass production and use are region- and site-specific. Therefore, as part of good governance of land use and rural development, bioenergy policies need to consider regional conditions and priorities along with the agricultural (crops and livestock) and forestry sectors. Bioenergy and new (perennial) cropping systems also offer opportunities to combine adaptation measures (e.g., soil protection, water retention and modernization of agriculture) with production of biomass resources.
- Several important bioenergy options (i.e., sugarcane ethanol production in Brazil, select waste-to-energy systems, efficient biomass cookstoves, biomass-based CHP) are competitive today and can provide important synergies with longer-term options. Lignocellulosic biofuels replacing gasoline, diesel and jet fuels, advanced bioelectricity options and biorefinery concepts can offer competitive deployment of bioenergy for the 2020–2030 timeframe. Combining biomass conversion with CCS raises the possibility of achieving negative carbon emissions in the long term—a necessity for substantial GHG emission reductions. Advanced biomaterials are promising as well for the economics of bioenergy production and mitigation, though the potential is less well understood as is the potential role of aquatic biomass (algae), which is highly uncertain.
- Rapidly changing policy contexts, recent market-based activities, the increasing support for advanced biorefineries and lignocellulosic biofuel options, and in particular the development of sustainability criteria and frameworks, all have the potential to drive bioenergy systems and their deployment in sustainable directions. Achieving this goal will require sustained investments that reduce costs of key technologies, improved biomass production and supply infrastructure, and implementation strategies that can gain public and political acceptance.

This is a mixed picture. On the one hand the potential, also in terms of sustainable deployment, is stressed. In addition, biomass plays an essential role in meeting ambitious GHG mitigation targets, in particular for transport fuels (including growing demand from aviation and shipping) and feedstocks for chemical industry. Opportunities achieving synergies with rural development and ecosystem services are additional benefits. But, achieving such an optimal outcome, comes with a large number of preconditions, most notably with respect land use & natural resources (Dornburg et al. 2010). Integral land use strategies (including both agriculture and biobased economy options, nature protection and rural development are required, which means combinations of policy priorities leading to complex governance questions. Furthermore, in many areas improved technologies and more experience is needed to improve environmental performance and economics, This is true for advanced technologies for fuels, biomaterials, power, carbon management (CCS), cropping systems in many different settings, further development and optimization of infrastructure, logistics, functional markets and chain of custody control. Last but not least, effective business models & investments are required.

It is by no means certain that such a mix of preconditions will be secured, despite the strong drivers to do so. Deployment of biomass may therefore remain fairly low at a 100 EJ level in 2050 as argued by the IPCC report as well. The development pathways that could be followed could be profoundly different going from region to region across the globe and over time (Lynd et al. 2011). Therefore, predicting future bioenergy markets and trade remains uncertain.

1.5 Key Priority Areas for Future Bioenergy Markets

Clearly, the strongly growing demand for biomass and biofuels make clear that there is a growing need to develop biomass resources and exploit biomass production potentials in a sustainable way and to understand what this means in different settings. In some markets, prices of biomass resources and fuels are volatile, including indirect effects on price of raw material prices for e.g. the forest industry as well as on food (e.g. sugar). Biomass markets are still immature and this is in particular true for the demand side of the market; many biomass markets, e.g. for solid fuels, rely on policy objectives and incentives, that prove to be erratic in many cases.

It is particularly important to develop both supply and demand for biomass and energy carriers derived from biomass in a balanced way and avoid distortions and instability that can threaten investments in biomass production, infrastructure and conversion capacity. Our understanding of how this is best organized and managed needs further improvement. International biomass markets have been mapped by Task 40, but to date available analyses, statistics and modelling exercises still have limitations.

Developing the sustainable and stable, international, bioenergy market is a long-term process. The Tasks aims to provide well aimed contributions to such (policy making) decisions in the coming years for market players, policy makers, international bodies as well as NGO's. Priorities include:

Biomass supplies To deliver refined insights in availability and potential production and supply of biomass resources on regional, national and global level. This explicitly includes a range of biomass residue streams, land use and competition for land and on various markets worldwide (i.e. including developing regions). This objective is in particular to be tackled by inter-task collaboration. Focus lays on development of supplies at large in relation to various drivers (demand development, improvements in production and logistics) and barriers (e.g. lack of investment, sustainability concerns).

Sustainability & certification To determine how the sustainability of biomass supplies, use and trade can be secured optimally and efficiently, in particular from a market perspective, with specific attention for the impacts of certification/verification on international biomass and biofuels trade.

Trade, market and demand dynamics To map and provide an integral overview of biomass markets and trade on global level, as well as for specific regions, identify and map new markets and products (such as Jatropha oil, demand from industry, heating markets, biorefining and future 2nd generation biofuels production). Improve the understanding on how biomass trade and markets respond to fluctuating (fossil) energy prices, developments on global markets for food and forestry products, emission trading and policies of different countries. Specific attention is paid to the balance between demand and supply aspects versus structures, institutions, drivers, technologies, etc. influencing demand and how to organize biomass trade under uncertainties observed in the market (see e.g. Verdonk et al. (2007) for a discussion on different possible governance models for international bioenergy markets).

Transport, logistics and trade Provide (further) insights in international biomass supply lines and logistic requirements (including new producing regions, i.e. developing countries and Eastern Europe) and how these can be optimised over time. This includes increasing our understanding on how costs of biomass production, pre-treatment and transport can be reduced over time, e.g. through better organisation and applying more efficient technology. Such work includes advanced forecasting analyses on required logistic capacity required to facilitate increased biomass use and trade.

1.6 Set-Up of the Book

This introductory chapter is followed by nine more chapters;

Chapters 2 and 3 respectively describe the global developments in liquid biofuels and solid biofuels trade over the past 10 years or so.

Chapter 4 gives a more detailed insight how important countries developed working biomass markets and engaged in either import or export of biomass and biofuels over time. Outcomes are compared and both specific national issues as well as common characteristics of biomass markets and trade are discussed.

Chapter 5 deals with the importance of logistics and supply chains in bioenergy systems, markets and trade and presents different logistic solutions and their performance in different settings.

Chapter 6 provides an overview of the sustainability debate around the use and trade of biomass and biofuels and subsequent development of sustainability frameworks and certification. Furthermore, implications for biomass trade and markets of those developments are assessed and discussed.

Chapter 7 discusses both drivers and barriers for further development of biomass markets and trade.

Chapter 8, deals with the expected and potential future developments in biomass demand, biomass supplies and possible implications for international biomass markets, based on a review and original analysis of global (energy) scenario's.

Chapter 9 discusses the role of business and the importance of investments for further (sustainable) development of biomass markets and trade.

Chapter 10 wraps up the book with a synthesis of findings and overall recommendations.

References

Chum, H., Faaij, A., Moreira, J., Berndes, G., Dhamija, P., Dong, H., Gabrielle, B. X., Goss Eng, A. M., Lucht, W., Mapako, M., Masera Cerutti, O., McIntyre, T. C., Minowa, T., Pingoud, K., Bain, R., Chiang, R., Dawe, D., Heath, G., Junginger, M., Patel, M., Yang, J. C., & Warner, E. (2011). Chapter 2, Bioenergy. In O. Edenhofer, R. P. Madruga, & Y. Sokona, et al. (Eds.), *The IPCC special report of the intergovernmental panel on climate change: Renewable energy sources and climate change mitigation*. New York: Cambridge University Press. Available at: http://srren.ipcc-wg3.de/report/IPCC_SRREN_Ch02.pdf

Dornburg, V., van Vuuren, D., van de Ven, G., Langeveld, H., Meeusen, M., Banse, M., van Oorschot, M., Ros, J., van den Born, G. J., Aiking, H., Londo, M., Mozaffarian, H., Verweij, P., Lysen, E., & Faaij, A. (2010). Bioenergy revisited: Key factors in global potentials of bioenergy. *Energy & Environmental Science, 3*, 258–267.

European Commission. (2009). Available at: http://ec.europa.eu/energy/renewables/biofuels/sustainability_criteria_en.htm

Faaij, A. P. C., & Domac, J. (2006). Emerging international bio-energy markets and opportunities for socio-economic development. *Energy for Sustainable Development* (Special Issue on Emerging International Bio-energy markets and opportunities for socio-economic development), *X*(1), 7–19.

FAO. (2008). *The state of food and agriculture 2008 – Biofuels: Prospects, risks, and opportunities*. Rome: Food and Agriculture Organization.

GBEP. (2008). *A review of the current state of bioenergy development in G8 + 5 countries*. Rome: Food and Agriculture Organization of the United Nations.

GEA. (2012). *Global energy assessment – Toward a sustainable future*. Cambridge/New York/Laxenburg: Cambridge University Press/International Institute for Applied Systems Analysis.

Heinimö, J., & Junginger, H. M. (2009). Production and trading of biomass for energy: An overview of the global status. *Biomass and Bioenergy, 33*(9), 1310–1320.

Hunt, S., Easterly, J., Faaij, A., Flavin, C., Freimuth, L., Fritsche, U., Laser, M., Lynd, L., Moreira, J., Pacca, S., Sawin, J. L., Sorkin, L., Stair, P., Szwarc, A., & Trindade, S. (2007). *Biofuels for transport: Global potential and implications for energy and agriculture* prepared by Worldwatch Institute, for the German Ministry of Food, Agriculture and Consumer Protection (BMELV) in coordination with the German Agency for Technical Cooperation (GTZ) and the German Agency of Renewable Resources (FNR). Oxford Sterling: Earthscan/James & James.

IEA. (2010a). *World energy outlook 2010*. Paris: International Energy Agency.

IEA. (2010b). IEA World Energy Outlook. Available at: http://www.worldenergyoutlook.org/

Jonker, G. J., Junginger, M., & Faaij, A. (2013). Carbon payback period and carbon offset parity point of wood pellet production in the Southeastern USA. *Global Change Biology – Bioenergy* (in press).

Lamers, P., Hamelinck, C., Junginger, M., & Faaij, A. (2011). International bioenergy trade—A review of past developments in the liquid biofuel market. *Renewable and Sustainable Energy Reviews, 15*(6), 2655–2676.

Lamers, P., Hamelinck, C., Junginger, M., & Faaij, A. (2012). Developments in international solid biofuel trade—An analysis of volumes, policies, and market factors. *Renewable and Sustainable Energy Reviews, 16*(5), 3176–3199.

Lamers, P., Junginger, M., Dymond, C., & Faaij, A. (2013). Damaged forests provide an opportunity to mitigate climate change. *Global Change Biology* (in press).

Lynd, L., Abdul Aziz, R., de Brito Cruz, C. H., Chimphango, A. F. A., Cortez, L. A. B., Faaij, A., Greene, N., Keller, M., Osseweijer, P., Richard, T., Sheehan, J., Chugh, A., van der Wielen, L., Woods, J., & van Zyl, E. (2011). A global conversation about energy from biomass: The continental conventions of the global sustainable bioenergy project. *Interface Focus, 1*(2), 271–279. doi:10.1098/rsfs.2010.0047.

Metz, B., Davidson, O. R., Bosch, P. R., Dave, R., & Meyer, L. A. (2007). *Climate change 2007: Mitigation contribution of Working Group III to the fourth assessment report of the*

Intergovernmental Panel on Climate Change. Cambridge/New York: Cambridge University Press.

REN21. (2010). *Renewables interactive map*. Renewable Energy Policy Network for the 21st century. Paris: REN21.

Sikkema, R., Steiner, M., Junginger, M., Hiegl, W., Hansen, M. T., & Faaij, A. (2011). The European wood pellet markets, current status and prospects for 2020. *Biofuels, Bioproducts and Biorefining, 5*(3), 250–278.

Smeets, E. M. W., Faaij, A. P. C., Lewandowski, I. M., & Turkenburg, W. C. (2007). A quickscan of global bio-energy potentials to 2050. *Progress in Energy and Combustion Science, 33*(1), 56–106.

van Dam, J., Junginger, M., & Faaij, A. P. C. (2010). From the global efforts on certification of bioenergy towards an integrated approach based on sustainable land use planning. *Renewable and Sustainable Energy Reviews, 14*(9), 2445–2472.

Verdonk, M., Dieperink, C., & Faaij, A. P. C. (2007). Governance of the emerging bio-energy markets. *Energy Policy, 35*(7), 3909–3924.

Wicke, B., Verweij, P., van Meijl, H., van Vuuren, D. P., & Faaij, A. P. C. (2012). Indirect land use change: Review of existing models and strategies for mitigation. *Biofuels, 3*(1), 87–100.

Chapter 2
Developments in International Liquid Biofuel Trade

Patrick Lamers, Frank Rosillo-Calle, Luc Pelkmans, and Carlo Hamelinck

Abstract This chapter describes the past developments, current status, and trends in global liquid biofuel production and trade. Apart from providing quantitative overviews, it also elaborates why markets developed as they did. By 2011, close to 2,500 PJ of liquid biofuels were produced globally; over two-third of which were fuel ethanol and the remaining biodiesel. The feedstock base is exclusively regionally specific oil, sugar, or starch crops. Global trade in biodiesel has been and will in the foreseeable future be primarily driven towards the European Union, where renewable energy policies stimulate the consumption of sustainable transport fuels – although the EU biofuels market growth is slowing down. Fuel ethanol is largely produced and consumed in the Americas, with the USA and Brazil dominating global production, trade and deployment. International trade is both supply and demand driven. National support policies increased the domestic market value of biofuels and shaped demand side developments. Trade flows emerged where such policies were not aligned with respective trade measures. Import duties had the strongest effect on trade volumes while trade routes were influenced by tariff preferences. Most trade regimes appear to have been designed and adapted unilaterally along national interests causing market disruptions, trade inefficiencies and disputes.

P. Lamers (✉)
Copernicus Institute, Utrecht University, Utrecht, The Netherlands
e-mail: p.lamers@uu.nl; pa.lamers@gmail.com

F. Rosillo-Calle
Imperial College, London, UK
e-mail: f.rosillo-calle@imperial.ac.uk

L. Pelkmans
VITO, Mol, Belgium
e-mail: luc.pelkmans@vito.be

C. Hamelinck
ECOFYS, Utrecht, The Netherlands
e-mail: c.hamelinck@ecofys.com

M. Junginger et al. (eds.), *International Bioenergy Trade: History, status & outlook on securing sustainable bioenergy supply, demand and markets*, Lecture Notes in Energy 17,
DOI 10.1007/978-94-007-6982-3_2,

2.1 Introduction, Objective, and Methodology

Few markets have seen a similar internationalization trend and turbulences due to policy regulations and changes over the past decade in the bioenergy field as liquid biofuels.[1] We have seen an exponential growth in global production and trade; with volumes being strongly linked to national policies (Lamers et al. 2011). Changes in these policy frameworks though have also shown how vulnerable markets and trade patterns still are. Biofuel markets are inherently linked to other sectors, agriculture in particular, and face significant market disturbances some of which have led to various inefficiencies in the past.

The main objective of this chapter is to describe the past developments and the current status of global biofuel production and trade. Apart from providing quantitative overviews, it also elaborates why markets developed as they did and provide a methodological assessment and understanding of the numerous influencing factors. The chapter closes with possible scenarios for the near future based on past experience.

The basis of the production data is derived from industry sources (e.g. ePURE, UNICA) and double-checked with government statistics (e.g. USDA, Eurostat) and other literature (e.g. REN21). Trade data was collected via international statistics (e.g. USDA GAIN, UN COMTRADE) but contains uncertainties linked to trade codes as no harmonized system (HS) classifications beyond digit-6-level apply to biofuels. Countries can define their own codes and definitions, e.g. minimum blend volumes. Specific trade data quantifications therefore depended on previous IEA T40 work (Walter et al. 2007; Rosillo-Calle et al. 2009) and scientific literature (Walter et al. 2008; Lamers et al. 2011), which provided anecdotal data and methodologies to account for energy related trade only.

Trade flows of fuel ethanol are hard to statistically distinguish from trade in denatured, non-denatured, and mixed ethanol blends as no HS classification by end-use exists. We follow the methodology by Lamers et al. 2011, after which the fuel ethanol imports of country A (e.g. Nigeria) from B (e.g. Brazil) in any given year are derived by: (1) taking the total ethanol imports by Nigeria (e.g. 10 PJ), (2) defining the relative share per import country (e.g. 5 PJ from Brazil i.e. 50 %), (3) subtracting the local Nigerian fuel ethanol production (2 PJ) from the local fuel ethanol consumption (8 PJ) and thus defining the net import demand for fuel ethanol (6 PJ). Hence, of all imported ethanol, 6 PJ will be for fuel. Assuming that ethanol is imported and stored independent of end-use, we can say that 50 % of the imported fuel ethanol stems from Brazil, i.e. 3 PJ.

2.2 Developments in Liquid Biofuel Production

While global liquid production has been relatively constant across the 1990s, it has seen an exponential growth over the past decade (Fig. 2.1). 50 % of all biofuels were produced in North America (largely the USA) by 2011. Fuel ethanol is the leading

[1] In the context of this chapter, we refer to liquid biofuels merely as biofuels.

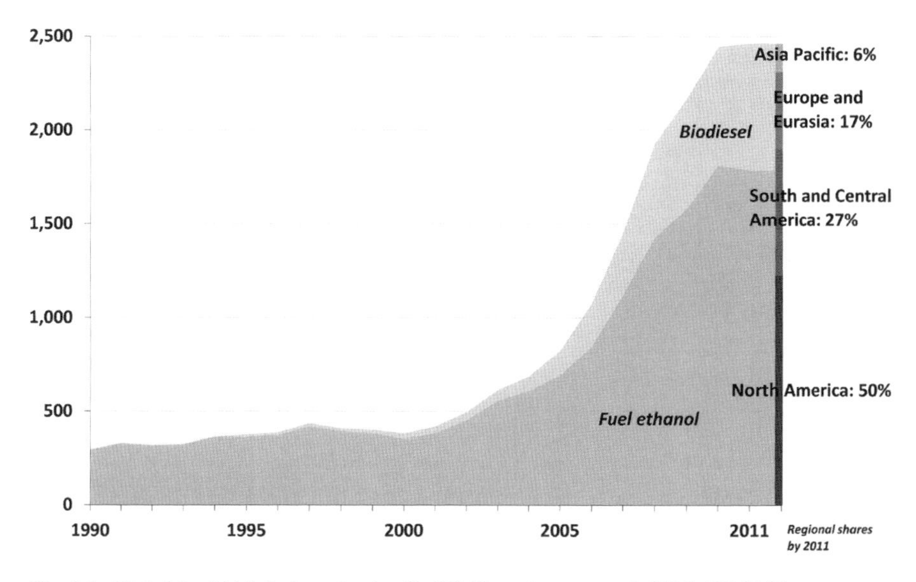

Fig. 2.1 Global liquid biofuel production [in PJ] (Data: Lamers et al. 2011; BP 2012)

biofuel at about 72.5 % of world production by 2011. It is almost exclusively produced across the Americas, with little production in Europe and Asia. World biodiesel production has been centred within the European Union (EU) and Asia, whereas recent production increases predominated in South America.

2.2.1 *Biodiesel Related Vegetable Oil Production*

The top vegetable oil markets are food (>80 %) and industrial applications (including biodiesel). Since biodiesel production did not really pick up significantly prior to 2008, the main driver for the vast expansion of vegetable oil production since the late 1990s (Fig. 2.2), palm and soya oil in particular, can be attributed exclusively to the growing demand for oils and high value protein in the food sector; particularly in emerging countries such as China and India caused by, among other things, population growth, improved standards of living and changing diets (Rosillo-Calle et al. 2009).

Biodiesel feedstock depends on the geographic region. Biodiesel in the US, Argentina, and Brazil, is almost exclusively soya oil methyl-ester (SME), whereas Indonesian and Malaysian biodiesel is palm oil methyl-ester (PME). In the EU it is traditionally rapeseed oil methyl-ester (RME) although shares of PME and SME have increased since 2005. Until a few years ago, the use of palm and soya oil in EU biodiesel was limited by technical requirements to the final product. Now, less strict technical requirements as well as the production of drop-in biodiesel (by hydrotreating vegetable oil) allow for even larger fractions of non-rapeseed oil. Expansion trends

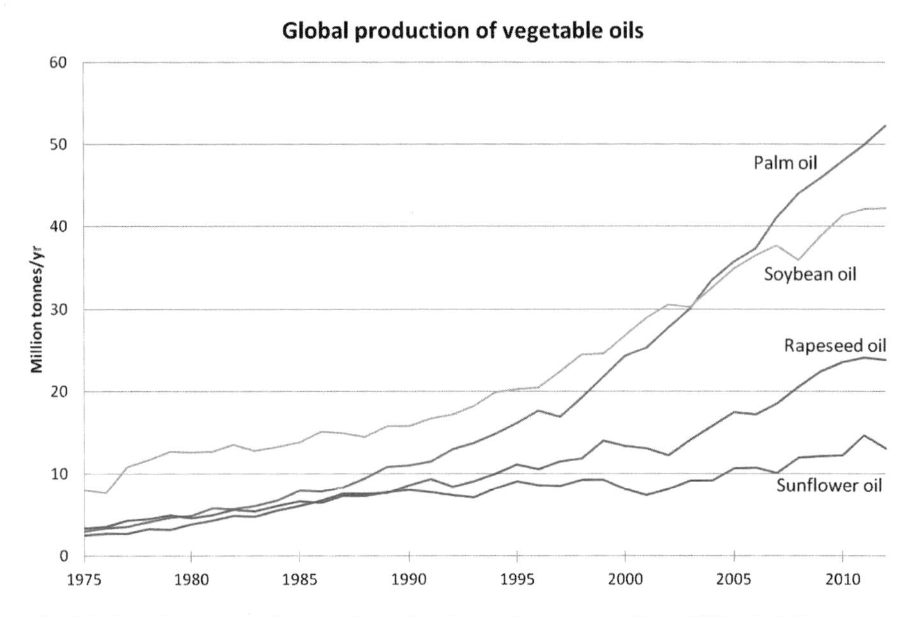

Fig. 2.2 World vegetable oil production (all end-use) [in Mtonnes] (Data: USDA 2012b)

of rapeseed oil production post 2003 show a strong link to the introduction of support policies for biodiesel in Europe. Biodiesel production increases in other world regions were much more recent and also linked to aforementioned feedstock, i.e. palm oil in Asia and soya oil in South America.

In recent years, we have observed important new actors and trends on the supply and demand side. China is rapidly emerging as the world leading importer of vegetable oils. Indonesia, Malaysia and Argentina dominate the export market of vegetable oil, representing approximately 75 %. Brazil has become one of the world's largest exporters of soybeans,[2] next to the USA. Also Argentina, whose share in this market has been continuously growing, is rapidly becoming one of the world's top exporters of soya oil.

The vegetable oil market is bound for major changes, and will face substantial challenges and opportunities. Improving living standards in emerging economies, population growth together with changing diets and the expansion of biodiesel, are new trends that will have a major impact in the future development of this sector.

2.2.2 *Biodiesel Production*

Biodiesel has been utilized as a substitute for mineral diesel since the early twentieth century, though in small quantities. Significant production increases only came after the European introduction of indicative biofuel targets in 2003 via the

[2] Soybeans are traded for animal feed (soybean meal) and soybean oil.

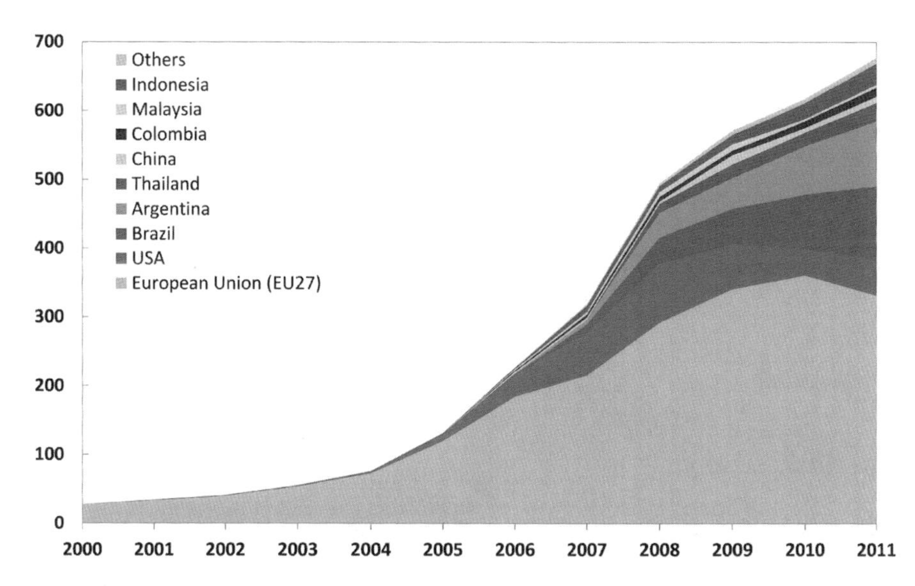

Fig. 2.3 World biodiesel production by country [in PJ] (Data: Lamers et al. 2011; BP 2012; Eurobserver 2012; Licht 2012; Lamers 2012)

EU-directive on the promotion of the use of biofuels or other renewable fuels for transport (2003/30/EC); triggering initiatives on EU Member State level to promote biofuels via tax exemptions or as a blend component in fossil fuels (more information in the Policy Sect. 2.5 of this chapter).

The EU, with its core production centres in Germany and France (followed by Spain, Italy, and Poland), has dominated biodiesel production globally over the past decade (Fig. 2.3). In recent years, this dominance has been challenged by increased generation in other world regions; though much of it can be linked to exports to the EU (more information in the Trade Sect. 2.3 of this chapter). An exception to this is Brazil whose production is merely consumed nationally. European biodiesel production is assumed to stabilize or even decline in the coming years, whereas further growth is expected in South America (Brazil, Argentina and Colombia) and Asia.

2.2.3 Fuel Ethanol Production

Close to 90 % of world fuel ethanol is produced in the US and Brazil which cover around two-thirds and one-third of this share respectively (Fig. 2.4). Overall volumes have increased over the past decade, but varied significantly with harvest qualities: the 2009 floods in the mid-west USA have reduced corn harvests and slowed ethanol production growth, the low sugarcane harvest around the world (and resulting high world sugar prices) dropped Brazilian fuel ethanol output in 2011. The export of Brazilian surplus ethanol was almost zero in the past 3 years.

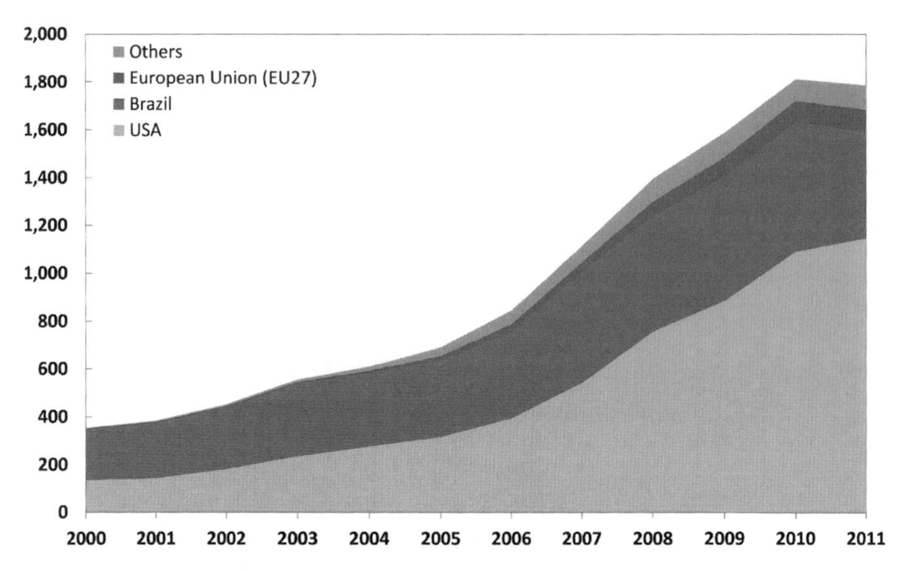

Fig. 2.4 Global fuel ethanol production [in PJ] (Data: Lamers et al. 2011; BP 2012; Eurobserver 2012; Licht 2012)

Significant production in other world regions is mostly concentrated in the EU where production has risen to 4.4 Billion Litres (93 PJ) in 2011 (Eurobserver 2012); representing 5 % of global production. China currently produces around 2 Billion Litres (Licht 2012).

2.2.3.1 Emerging Markets in Africa

Africa, with vast areas of idle land, an underdeveloped agricultural sector, cheap labour and favourable import conditions to the EU has an enormous potential in bioenergy feedstock production. Especially in sub-Saharan Africa, there have been many biofuel initiatives in the past 5 years. In reality, most of these initiatives failed or remained small-scale. Today the African biofuel production is still practically non-existent, let alone biofuel export to the rest of the world. Table 2.1 below shows the total production of biodiesel and ethanol in Africa in 2010. In Malawi, biofuels have been promoted since the 1980s, in Zimbabwe the production of biofuels is promoted since 2004. The large ethanol production in Sudan results from one existing 100,000 ha sugar cane plantation in Kenana which aims to export fuel ethanol to the EU market. Total African biofuel production combined is so far smaller than the average output of one European facility. There are only a handful of serious large-scale new projects that focus on producing for the EU market (e.g. in Sierra Leone and Tanzania).

Nevertheless, there are sincere concerns about socio-economic impacts of large-scale biofuel feedstock production in Africa, most importantly land-use rights and the loss of access to land and water resources. As impacts are partially related to the

Table 2.1 Biofuel production in Africa in 2010 [in ktonnes] (EIA 2012)

	Biodiesel	Ethanol
Sudan		22.8
Malawi		9.1
Zimbabwe	5.2	0.9
Ethiopia		4.6
South Africa	1.6	
Mozambique	1.0	
Tanzania	0.5	
Rwanda	0.5	
Total	**8.8**	**37.4**

establishment of plantations, the currently still small production is a poor indicator. Several biofuel projects developed over the past 5 years indeed include aspects of land grabbing; though as analysis shows (e.g. Ecofys 2012), not at the scale that has often been suggested.

Governments of several African countries, such as Tanzania and Mozambique have noted the importance of guiding the biofuels development towards benefits for their countries. Mozambique adopted a biofuels strategy in 2009, to reduce the country's dependence on oil imports and to strengthen agricultural development and food security. In Tanzania, a holistic biofuels policy framework is under development, covering not just the production and deployment of biofuels, but also requesting that the domestic market will be fulfilled before biofuels are exported, and putting much focus on socio-economic aspects.

2.2.3.2 Emerging Markets in Asia

India is currently the world's second largest producer of sugarcane (after Brazil), mainly used for sugar production to satisfy its huge domestic market. The Government's National Policy on Biofuels, adopted in 2009, proposes a target of 20 % blending of biodiesel and bioethanol by 2017. India's biofuel strategy continues to focus on the use of non-food resources; namely sugar molasses for production of ethanol and non-edible oils for the production of biodiesel. The target is to blend 5 % ethanol to petrol. It has however not been achieved due to the short supply of sugar molasses in 2008–2010 harvest seasons and the overall low sugarcane crop production. Ethanol is a major feedstock in the chemical industry in India. Unlike China, India does not seem to have any clear policy on import/exports. Currently, India has 330 distilleries producing around 4 billion litres of rectified spirit (alcohol) annually. About half of those distilleries have the capacity to produce 1.8 billion litres of conventional ethanol per year, sufficient to meet the 5 % blending mandate (USDA 2010b).

China is the world's third largest producer of sugarcane ethanol, although cassava, wheat and maize are also used. Ethanol production almost tripled in the last 10 years, reaching over 2 billion liters by 2010 (Licht 2012). The current use of biofuels in China is part of a strategy to decrease oil imports, foster agricultural and social development, and promote environmental sustainability. Currently the situation is delicate due to a combination of poor harvests and food price hikes causing concern regarding potential ethanol production expansion in the country. China is currently a net-importer of ethanol; a situation which would be fostered should the government decide to expand the use of ethanol fuel.

2.3 Developments in International Trade 2005–2011

2.3.1 Biodiesel

Global net biodiesel trade has grown from practically zero in 2005 to almost 2,500 ktonnes by 2011. Trade almost exclusively targets blending and consumption within the EU; only a marginal fraction is traded elsewhere, e.g. EU exports to Norway. Today's key export nations include Argentina, Indonesia, and the US; although the past years have shown dramatic changes in trade routes and volumes due to support policy changes in the US and the introduction of EU trade measures. Malaysian exports and trade volumes have come to a halt by 2010; apparently linked to the absence of local policy support and strong competition from (supported) Indonesian biodiesel. Brazil, despite its large production, has so far remained a closed market; primarily due to remote plant locations and relatively high production prices in comparison to other exporters.

Figure 2.5 shows a timeline of biodiesel trade flows between 2008 and 2009, the key phase of the splash-and-dash era; and the current situation as by 2011. It highlights that early trade to the EU came primarily from and via the US. This was largely driven by the US volumetric excise tax credit (VETC) given to biodiesel blended with fossil fuel; established in 2004. As the VETC was not linked to domestic production or consumption, it could also be collected for imports/exports. This allowed and ultimately established a practice called 'splash-and-dash', under which biodiesel was shipped to the US from a third country (e.g. from Argentina) solely to claim the tax credit before it was re-exported[3] again. These re-exports exclusively went to the EU where the biodiesel would receive a second financial incentive through many Member State's support schemes (Lamers et al. 2011). Another term, 'B99 effect', was linked to the same phenomenon as the VETC definition of blending made it possible to receive the credit by adding <1 % of mineral oil resulting in trade of B99 biodiesel. By March 2009, the EU put anti-dumping and countervailing duties on US biodiesel imports in place. This dramatically reduced US trade volume, and clearly revealed their original origins, as biodiesel was again traded directly towards the EU, e.g. from Argentina, Malaysia, Indonesia (Fig. 2.5). The

[3] Re-exports are defined as exports of previously imported commodities.

introduction of trade measures quickly lead to increasing triangular trade, i.e. US produced biodiesel was shipped to the EU via Canada (in 2009) and more recently also India. Canada was ultimately included in the anti-dumping and countervailing duty regulation, whereas EU-investigations to include, e.g. Singapore, the prime hub for palm oil derived biodiesel, could not verify similar practices.

Obviously, import volumes differ between EU Member States in terms of EU-internal and international trade (see Lamers et al. 2011; Eurostat 2012, 2013 for details). Biodiesel supply in the Netherlands, the UK, Spain, Portugal, and Italy is covered to a large extent by EU-external imports whereas those are

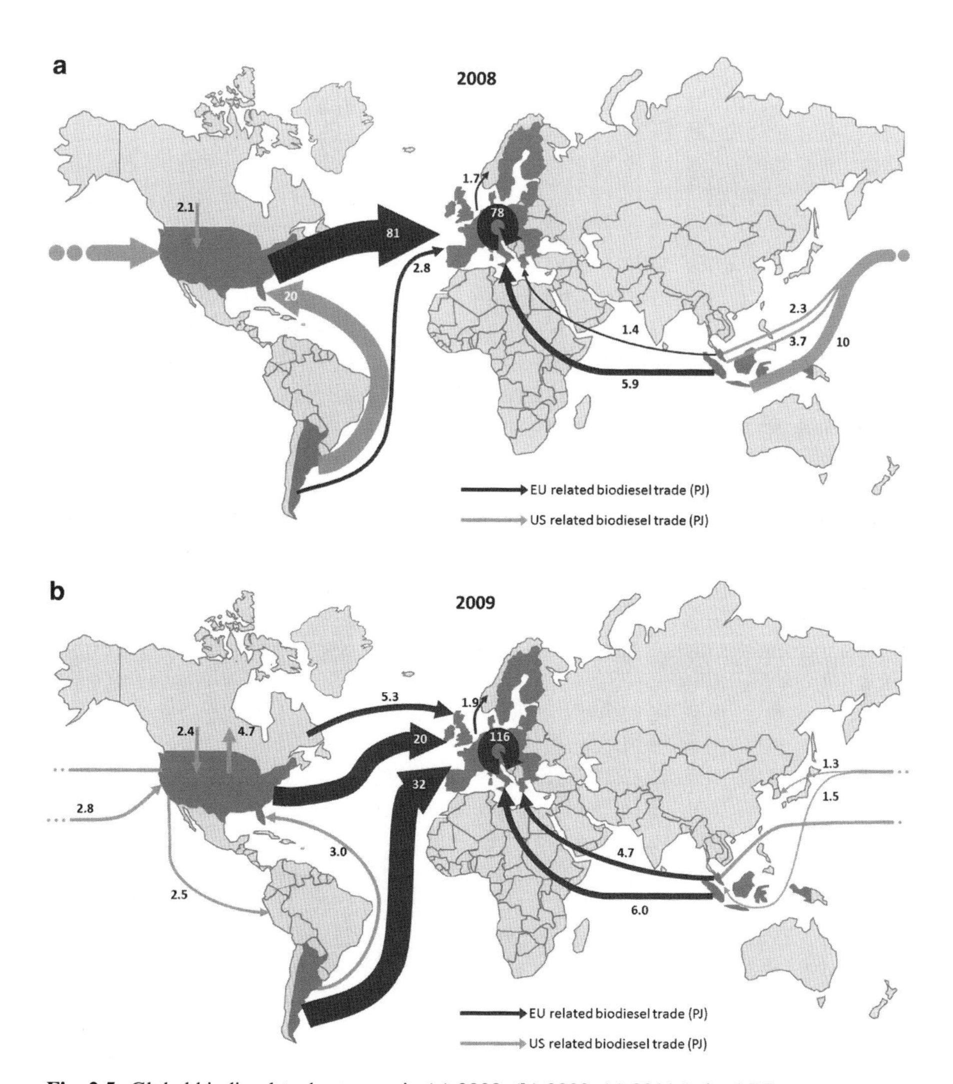

Fig. 2.5 Global biodiesel trade streams in (**a**) 2008, (**b**) 2009, (**c**) 2011 (min. 1 PJ)

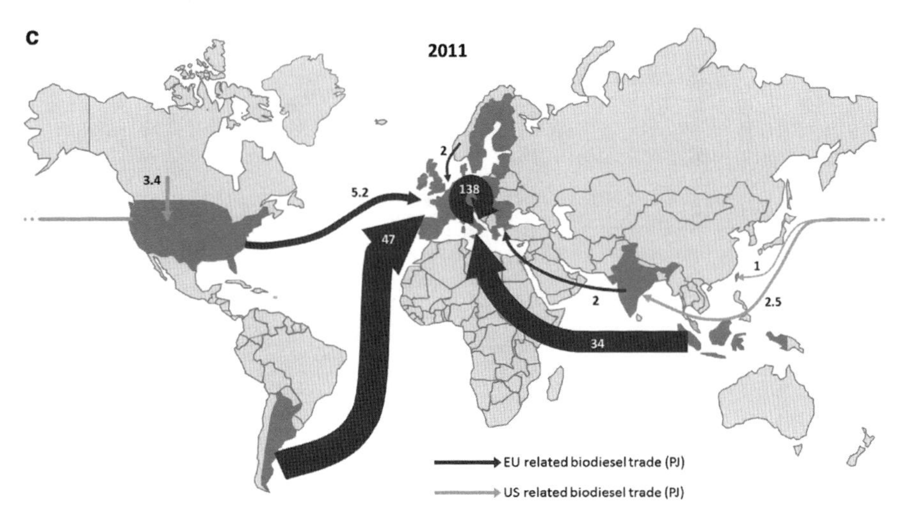

Fig. 2.5 (continued)

marginal, e.g. in the German biodiesel trade balance (Eurostat 2012, 2013; Lamers 2012). Existing oilseed crushing capacity and production cost reductions via feedstock imports (e.g. rapeseed from Poland) are assumed to have safeguarded German biodiesel production against competition from imports so far. The Netherlands, due to its large fossil fuel refining capacities in Rotterdam and Amsterdam, is the largest biofuel distribution place in Europe. In recent years, imports from Argentina at prices below production costs in Spain and Portugal have troubled the biodiesel industry in these markets. By summer 2012, the Spanish government issued a new biofuel policy aiming to ban imports of Argentinean and Indonesian biodiesel. Argentina has filed an official complaint at the WTO on this aspect.

In addition, the introduction of minimum sustainability criteria for liquid biofuels under the EU Renewable Energy Directive (RED) 2009/28/EC has led to complaints by key exporting nations, primarily Indonesia and Argentina; although so far without WTO sanctions or other trade implications. It is also reported that since mid-2011, certificates awarded to RED compliant rapeseed-methyl-ester (RME) were transferred to biodiesel volumes made from other, non-RED feedstock to enter the German market (AM 2012). While the practice has eventually been officially acknowledged by Germany's finance ministry, the German industry association is currently trying to reverse this decision, claiming it undermines demand for European RME while artificially swelling consumption of RED-compliant biodiesel for summer grades made from imported palm and soya oil methyl-ester.

2.3.2 Fuel Ethanol

Fuel ethanol is traded together with ethanol for other end-uses. A differentiation is inherently difficult to make, especially since trade codes do not specify end-use (see Methodology Section for details). Limiting trade streams to markets where respective policies have stimulated fuel ethanol consumption in transport, the main routes are between the EU, US, and Brazil (Fig. 2.6).

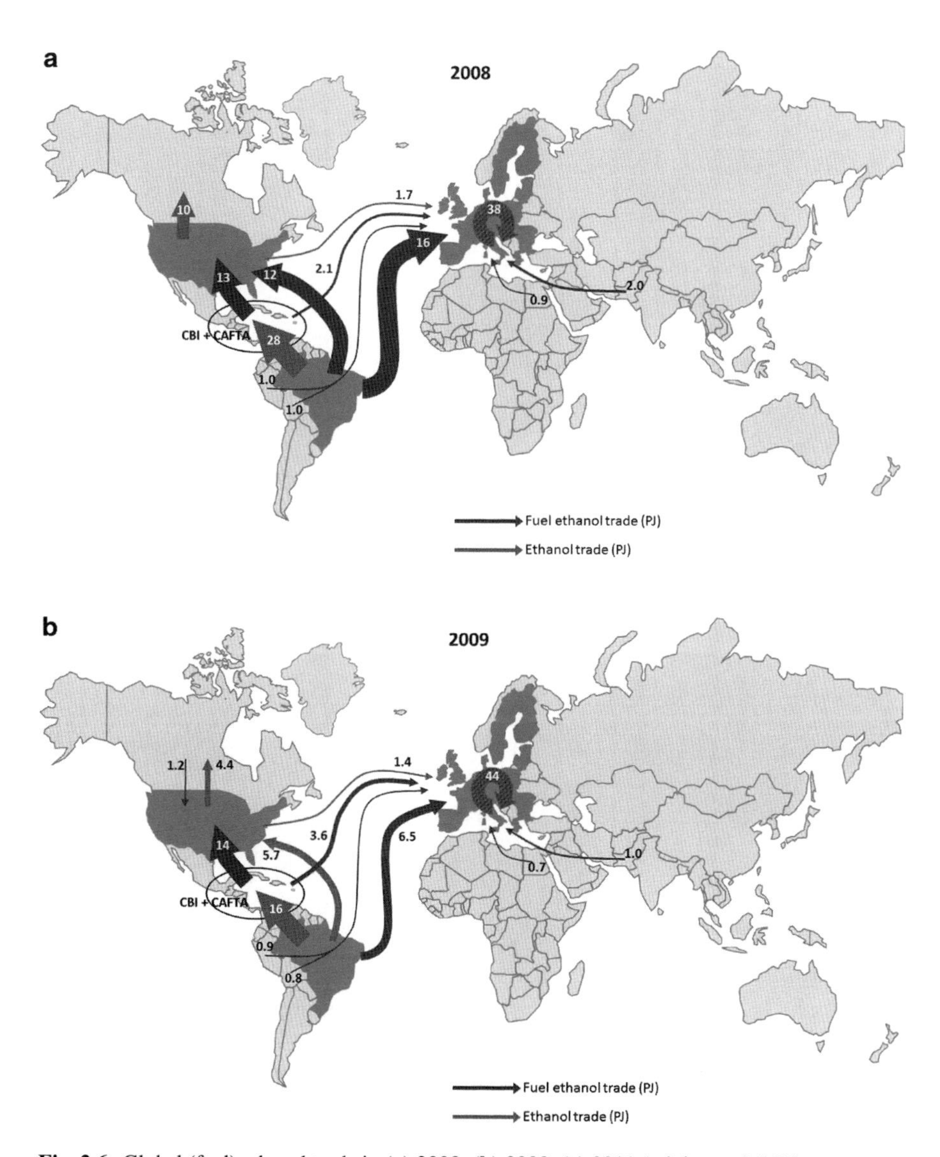

Fig. 2.6 Global (fuel) ethanol trade in (**a**) 2008, (**b**) 2009, (**c**) 2011 (minimum 0.5 PJ)

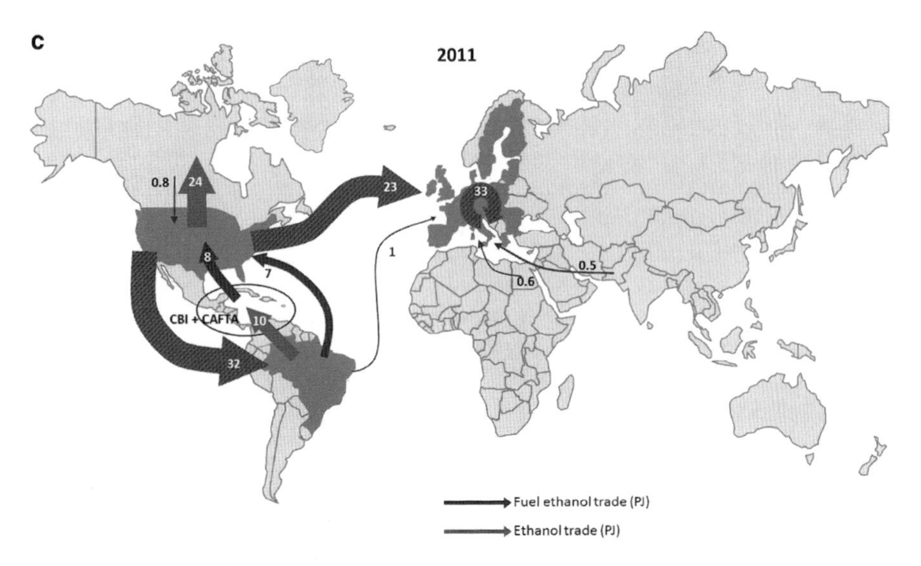

Fig. 2.6 (continued)

For most of the past decade, Brazil has been the leading export nation of (fuel) ethanol. Its highly efficient sugarcane production and conversion made it highly price competitive in many markets, including the US and the EU; where local markets were protected via respective tariff lines. In terms of export volume, Brazil was surpassed in 2011 by the US, whose strong production increase, high corn harvests, and a limited national market (E10) boosted export volumes; in particular to the EU. The US even exported ethanol to Brazil, whose low sugarcane harvest in 2011 made it a net importer.

Historically, with the introduction of the US Volumetric-Excise-Tax-Credit (VETC) in 2004, synthetic ethanol imports from Saudi Arabia were replaced by Brazilian imports as the previous were not eligible. The introduction of the VETC also increased the US market value for imported ethanol and it became economically viable to import Brazilian ethanol despite US import tariffs. In recent years, Brazilian ethanol has been increasingly transferred to the US via the Caribbean Basin (which enjoys tariff preferences). A marginal share has also originated in Canada (Fig. 2.6). US ethanol exports (all purposes) have mainly been destined for Canada and the EU (USDA 2012a). Previous exports to Mexico were diverted towards the EU after the introduction of the EU biofuels quota.

Most of the EU's ethanol imports originated in Brazil, while shares from nations subject to tariff preferences in particular Central and South America, increased until 2009. As of 2010, imports of US corn based ethanol increased sharply. In many EU Member States only undenatured, i.e. drinkable, ethanol is eligible to fulfill the respective nation's biofuel quotas (USDA 2009, 2010a). This effectively works as a trade protection against lower priced imports (mainly from the US and Brazil), as EU tariffs for undenatured ethanol (0.192 €/l) are almost twice as high as those of denatured ethanol (0.102 €/l).

It is not surprising that many efforts have been made to circumvent EU ethanol tariffs. Most have aimed at importing under alternative tariff lines (with lower duties); an effect triggered by the absence of specific fuel ethanol custom codifications. A relatively prominent example was the so-called 'Swedish loophole' under which, by mixing ethanol with 12.5–20 % gasoline just prior to customs declaration, ethanol for fuel blending was imported into Sweden as a chemical compound under the 'other chemicals' tariff line (CN 3824); eventually reducing the tariff to 6.5 % rather than 0.192 €/l (equaling about 63 %) for undenatured or 0.102 €/l (equaling roughly 39 %) for denatured ethanol (Kutas et al. 2007; Lamers et al. 2011). Eventually legislative changes were made in 2007 to close the loophole.

In 2010, concerns were again raised that a similar practice was applied by US ethanol exports to the EU (REA 2010). The EU ethanol industry became aware of it as US imports rose to over 1.1 billion liters (23 PJ) by the end of 2011 (Fig. 2.6); the majority of which entered the UK, Netherlands, and Finland classified as a chemical compound (CN 3824.90.97) in the form of E90 (ePURE 2012; Eurobserver 2012). A new EU customs regulation was eventually put in place in spring 2012 which forces imports of fuel blends containing at least 70 % ethanol to be classified as denatured ethanol.

2.4 Links Between Agricultural and Liquid Biofuel Markets

Current ("1st generation") liquid biofuels almost totally rely on agricultural crops, with sugarcane, corn, wheat and sugar beet for bioethanol, and rapeseed oil, soybean oil and palm oil for biodiesel. This creates a strong link between the agricultural and liquid biofuel markets.

2.4.1 Evolution of Commodity Prices

Agricultural commodity prices have varied tremendously in the past years. The following Fig. 2.7 shows the price indices for cereals, oils and sugar between 2002 and 2012, on a monthly basis (source: FAO 2012).

For comparison, Fig. 2.8 below shows the evolution of crude oil prices, also on a monthly basis (source: EIA 2012).

It is quite remarkable that the indices of grains and vegetable oil have similar tendencies as crude oil prices, with a spike in 2008, a fall-back late 2008, early 2009, and a steady recovery by 2011. The sugar prices seem to have their own mechanisms, with high fluctuations over the past 6 years.

The spike of commodity prices in 2007–2008 has triggered several studies to investigate the reasons behind the increase in commodity prices. Biofuels were often blamed – particularly by NGOs and the media – to be the main reason for these increases, but looking back it is clear that several causes have played a role at the same time (Pelkmans et al. 2009):

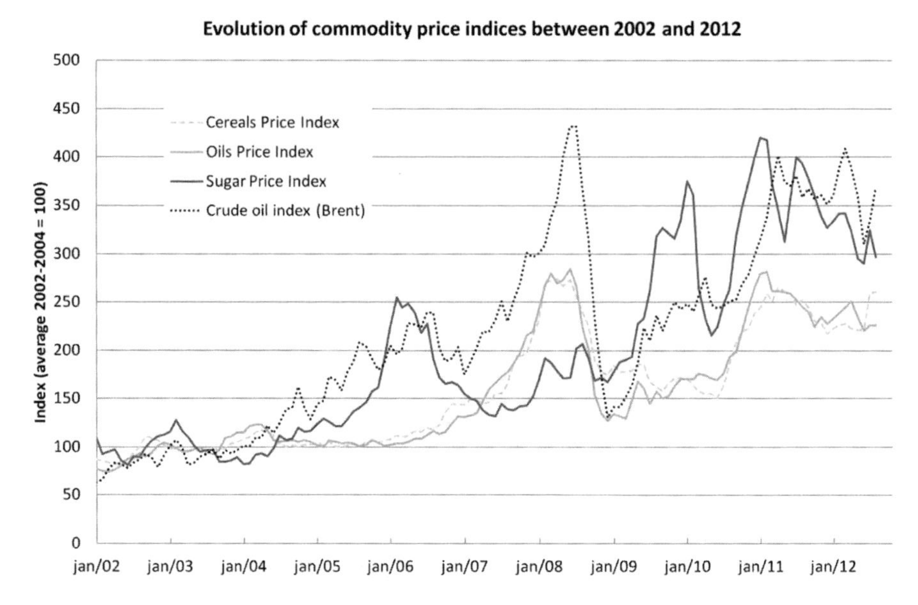

Fig. 2.7 Evolution of commodity price indices between 2002 and 2012 (Data: FAO 2012)

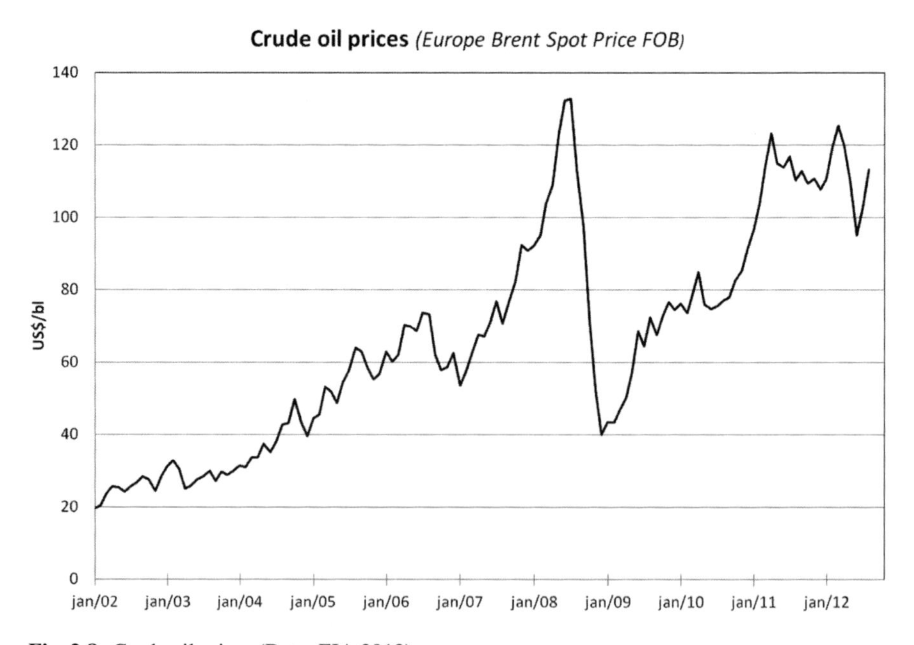

Fig. 2.8 Crude oil prices (Data: EIA 2012)

- increase of crude oil prices from 50 to over 140 $US per barrel,
- decrease of the value of the US dollar, as most markets are traded in this currency,
- speculation by the financial sector in agricultural commodities ("self-fulfilling prophecy" of increasing prices). This also seems to be linked to the low value of the US dollar,
- export restrictions in certain countries (e.g. Russia) as a response to expected global shortages,
- growing economies in Asia, with increasing demand for energy and food, and changing diets (e.g. increased consumption to meat),
- low crop yields in certain regions due to bad weather circumstances (e.g. Australia in 2006–2007 whose grain harvest fell by more than 50 %),
- decrease of stocks in the past years (not controlled by governments anymore, but by commercial parties who have interest in increasing prices),
- growing demand for biofuels.

In fact feedstock prices have increased much more than biofuel prices – for which the biofuel sector's financial margins were dropping – so the biofuel sector was apparently not the price-setter.

Since summer 2008, commodity prices have dropped again. Again different causes have played a role, but now in the other direction: the worldwide financial crisis has lowered energy demand and economic growth rates, reducing prices of crude oil and other commodities; crop production was again at normal level (while 2006 and 2007 were exceptionally bad, especially in Australia), so stocks could be filled again; speculative markets have reduced with the financial crisis. Meanwhile prices have recovered again, and are again in the same range as during the 2007–2008 period.

2.4.2 *Biofuel Feedstocks in Relation to Production and Trade Volumes of Agricultural Commodities*

We looked at typical cases where biofuel markets had possible interference with food and feed markets. Cases considered are corn, wheat and sugarcane for ethanol; rapeseed, palm oil and soybean oil for biodiesel.

2.4.2.1 Corn for Ethanol in the United States

The US accounts for roughly 30–40 % of global corn production, and is traditionally the world's dominant corn exporter (over 50 % of global corn trade), followed by Argentina and Brazil. While the US dominates world corn trade,

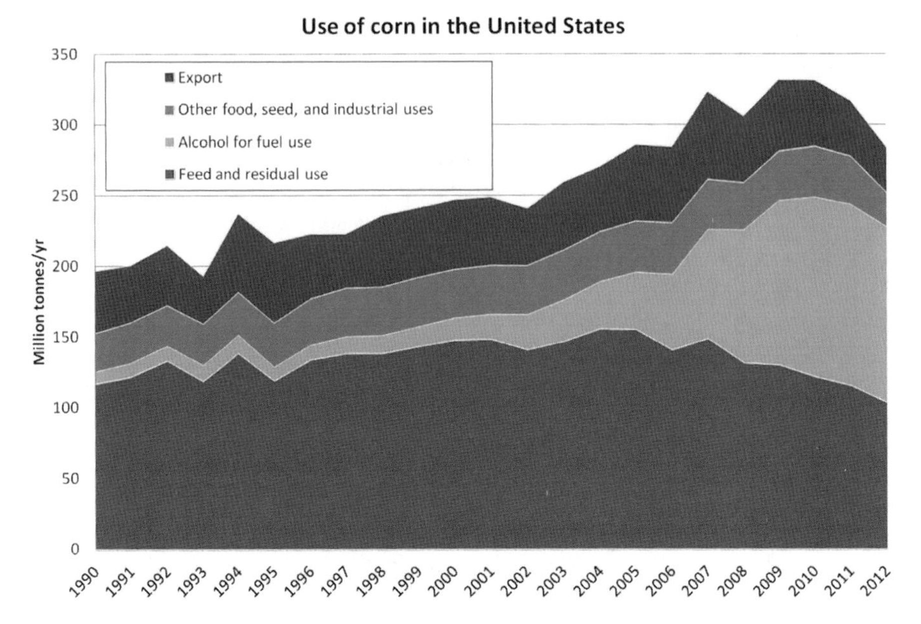

Fig. 2.9 Use of corn in the USA (Data: USDA 2012b)

exports only account for a relatively small portion of US corn use (about 15 %). This means that corn prices are largely determined by supply and demand relationships in the US market, and the rest of the world must adjust to prevailing US prices. As a result, the amount of corn grown in the US and the share of corn used for domestic consumption versus exports, has significant impact on international corn prices.

Corn use for ethanol represented 40 % of US corn production in 2011. The availability of US corn for feed, food and export markets has diminished since 2007 (Fig. 2.9). Mind that DDGS, the by-product of ethanol production, is now also available for the feed market.

2.4.2.2 Corn and Wheat for Ethanol in China and the EU

In the rest of the world (outside the US), the use of grains for ethanol is limited. EU ethanol is mostly based on wheat, but only few percent of European wheat production (2.6 % in 2008) is used for ethanol production. This has an insignificant effect on the availability of EU wheat for food, feed or export markets as most of the increase is covered by yield increases and extra land availability in East Europe.

In China 1.5 % of the local grain production (mainly corn) was used for ethanol in 2007–2008. While this number is also marginal, in response to high food prices, the government in 2007 suspended new ethanol projects based on edible grains, including any plans to expand existing plants.

2.4.2.3 Sugarcane for Ethanol in Brazil

The Brazilian sugarcane harvest is used almost evenly as feedstock for two major commodities: sugar and ethanol. Sugar and sugarcane prices have seen a high increase in 2010–2011, mostly due to reduced sugarcane harvests in Brazil related to bad weather conditions. As a consequence, more sugarcane was being refined into sugar and less into ethanol. This has led to a serious reduction of ethanol exports from Brazil. While Brazil has been exporting 15–20 %[4] of its ethanol production between 2005 and 2009, in 2010–2011 Brazil even became a net importer of ethanol from the US. Currently 8 million hectares of land is used to produce sugarcane. Brazil has no fundamental feedstock problems as it has ample space to extend its sugarcane production (outside rainforest regions). Nevertheless there are some concerns for this expansion. The expansion could happen on degraded grass planes, but there is a risk that fields in the natural Cerrado area or surroundings could be claimed for sugarcane expansion. Furthermore there could be indirect effect that extensive livestock breeding would shift to the north.

In the past, there seemed to be a price link between sugar, ethanol and petrol. However since 2006, sugar prices have behaved differently from the crude oil prices and the link is less pronounced. On the contrary, the price of sugar determines the price of ethanol on world markets.

2.4.2.4 Vegetable Oil for Biodiesel

In the past decades there has been a steady growth in the use of vegetable oils, with a prominent role for palm oil, soybean oil and rapeseed oil. While there is a growing role for biofuel use, consumption of vegetable oil for food keeps growing at a rate of 3–4 % per year. When comparing the growth of vegetable oil consumption from 2005 to 2011, vegetable oils for food have increased with more than 18 million tonnes, vegetable oils for biodiesel with around 13 million tonnes, while other industrial use is rather stable. Worldwide about 12 % of worldwide vegetable oil production is used to produce biodiesel (Fig. 2.10).

When looking at the division between palm oil, soybean oil and rapeseed oil, and its use in industrial vs. food use, the period around 2005 created a shift to more industrial applications, in all three cases. For soybean and rapeseed oil this can be directly linked to biodiesel production as their industrial application was rather modest before (Fig. 2.11).

2.4.2.5 Biodiesel in Europe

Europe has been the main player in biodiesel for a long time. Biodiesel was promoted in the 1990s, mostly to offer alternative outlets for agriculture, which was facing overproduction at that time. Traditionally, biodiesel in Europe is

[4] http://ageconsearch.umn.edu/bitstream/60895/2/Crago_CostofCornandSugarcaneEthanol_AAEA.pdf [January 2013].

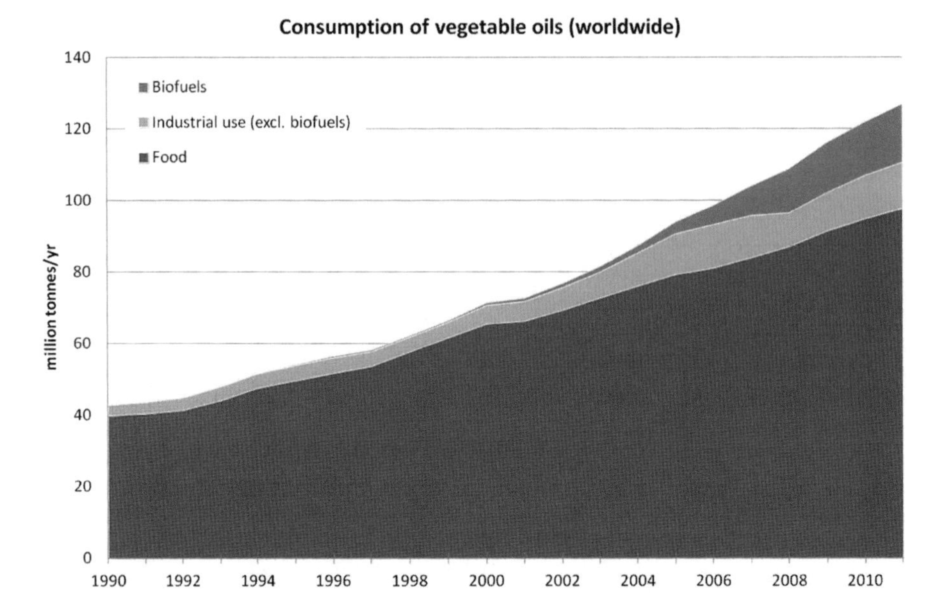

Fig. 2.10 Worldwide consumption of vegetable oils (Data: USDA 2012a, b, Licht 2012)

predominantly produced from domestically grown rapeseed. According to the Biofuels Baseline 2008 Study (Hamelinck et al. 2011), 66 % of biodiesel produced in the EU came from rapeseed, 13 % from soybean oil and 12 % from palm oil in 2008; the latter two are mainly imported. The reason for the dominant role of rapeseed oil is to be found in the tradition of producing rapeseed, its technical properties, and the high level of public support provided in EU countries. The increasing demand from the biodiesel sector has tightened the EU's vegetable oil balance, making feedstock imports for biodiesel production necessary. With the discussion on sustainability of biofuels, and increasing prices of biodiesel feedstocks, the biodiesel market in Europe has stagnated in the past years.

2.4.2.6 Biodiesel in North and South America

Only after 2005 other world regions started to introduce biodiesel in their diesel markets. Until 2005 industrial use of soybean oil was marginal, but from 2005 its use for biodiesel is growing, mainly in the USA and South America. Soybean oil use for food still grows at the same time. Biodiesel producers in South America benefit from a large exportable soya oil surplus (connected to soymeal production), part of it is also targeting export to the European market.

While soybeans are not the most efficient crop solely for the production of biodiesel, their common production and use for food products has led to soybean biodiesel becoming the primary source for biodiesel in the US.

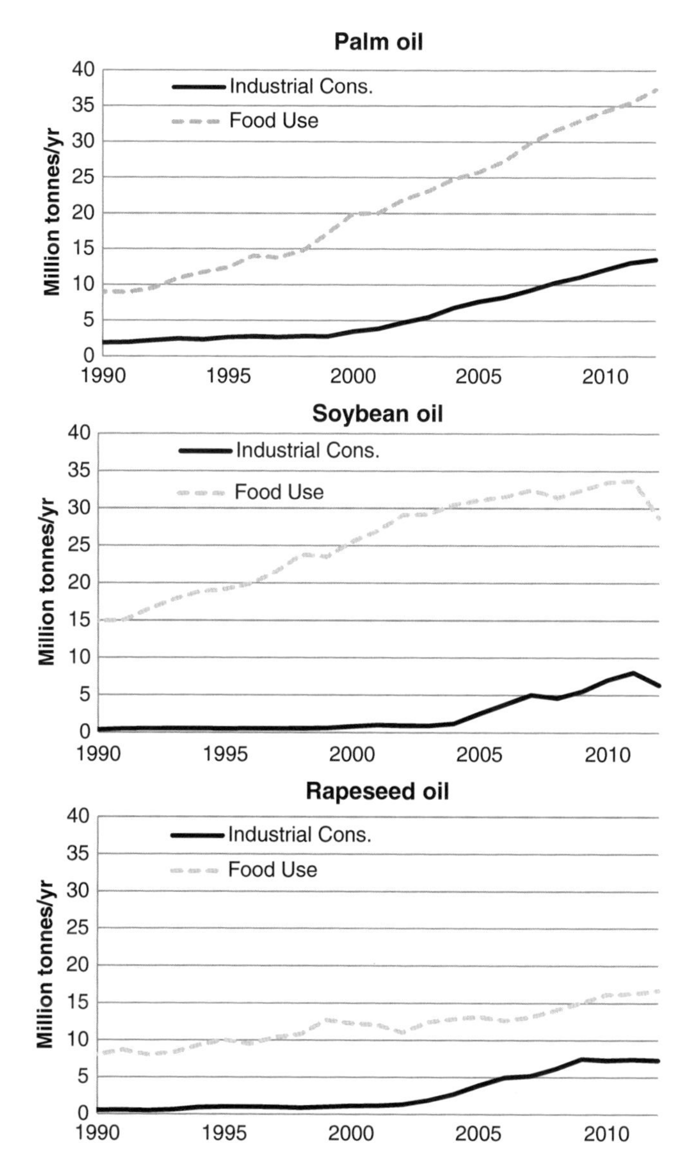

Fig. 2.11 Global consumption of palm, soybean, and rapeseed oil (Data: USDA 2012a, b)

2.4.2.7 Biodiesel in South-East Asia

Around 80–85 % of worldwide palm oil is produced in Indonesia and Malaysia, and most of it is exported to the rest of the world for food purposes. The global use of palm oil for food has actually doubled in the past 8 years. Since 2003 industrial applications are also growing; this may be partly related to biodiesel production, partly to other oleochemical applications (possibly replacing other vegetable oils for these applications). While the role of palm oil for biodiesel production has been limited so far, also from a quality point of view (low temperature behaviour), the instalment of HVO technology (hydrotreated vegetable oil), as e.g. the NExBTL facilities in Finland, Rotterdam and Singapore creates possibilities to enlarge the share of palm oil for biodiesel. On the other hand sustainability requirements for the European market might restrict the application of palm oil for biodiesel in the future.

The previous discussion shows that biofuels are indeed taking a substantial share of some agricultural commodities on global level, in particular for vegetable oils, corn and sugar. While there is already a link between commodity prices and energy prices, this may be reinforced through the applications of (1st generation) biofuels. Thus, if crude oil prices remain high, in the long run, biofuel feedstock prices will experience an upward pressure as well. If we continue to rely on biofuel feedstocks that are used directly to produce food or that are produced on land that would be producing food, then we will strengthen the direct link between crude oil prices and food prices. There may be some disagreement about the magnitude of the impact on food prices from biofuels, but there is no disagreement that there is an impact.

2.5 Policies and Other Influencing Factors

Biofuels were promoted by governments worldwide for a number of reasons, e.g. reducing greenhouse gas (GHG) emissions, enhancing the security of energy supply, but also revenue generation for local industry and job creation (the US VETC e.g. was part of the US Job Creation Act; USCH 2004).

Moreover, as past analysis shows, EU and US biofuel policies, originally strictly aimed at promoting domestic industry, had significant impacts on world biofuel production and trade patterns (Lamers et al. 2011). The main reasons for unintended impacts on international trade seem to be the mere focus on steering domestic production and consumption while neglecting international trade aspects (market factors) in policy making. A clear example on this is the design of the US VETC. The US fossil fuel consumption in passenger transport is mainly petrol, i.e. fuel ethanol is the key biofuel. To prevent ethanol imports from being blended under the VETC, i.e. to favour local production, the US levied an import tax on fuel ethanol. Such a tax though was not put on biodiesel imports allowing for trade effects as observed under the ‘splash-and-dash practice’. Similar examples exist also for the European context, where the phase-out of tax exemptions and the introduction of blending mandates across several EU Member States led to import increases as blenders preferred cheaper imports (PME and SME) over domestically produced RME. The second EU example was the introduction of anti-dumping and countervailing measures against US produced biodiesel that neglected the option of triangular trade or the general possibility for traders to down-blend and import biodiesel below the EU customs mark of B20 concentrations (under CN 3824.90.91).

While biofuel support policies in the EU and the US have prompted an increased international production and trade in liquid biofuels, it is important to stress that actual trade flows evolved due to interconnected and additional market/economic factors. It was these market factors, i.e. price differences connected to support policies which directed international biofuel trade flows towards one region or the other. Generally, support policies (artificially) increased the domestic market value for biofuels. Wherever such policies/prices were not accompanied by trade measures restricting trade volumes or imposing import duties, international trade developed (Lamers et al. 2011). However, even under the presence of trade measures, trade was economically viable for export regions with large resource potential and relatively low production costs compared to the destination markets (i.e. Argentinean, Indonesia, or US exports to the EU).

International trade in liquid biofuels is both demand and supply driven. Production costs and trading options are also influenced by additional short- and long-term market factors such as varying international feedstock and crude oil prices. A complete overview of influencing factors is given in Table 2.2. To steer international biofuel trade, policy makers would need to influence the economics of trade (see Kaditi 2009 for a broader discussion).

2.6 Outlook and Conclusions

International biofuel markets have grown exponentially over the past decade. Today's markets, although still volatile and policy dependent, have become much more transparent. The biodiesel industry has become interwoven with the already existing global vegetable oil and oilseed market (players). Trade volumes have increased from practically zero 10 years ago, to 2.5 Mtonnes (94 PJ) of biodiesel by 2011. The EU has been and will most likely remain the key production and consumption region for biodiesel until 2020. Many countries have followed suit and implemented national blending targets for biodiesel, thus stimulating domestic production and consumption. Partly, their production has been targeted for export to the EU. Such trade is expected to grow in the future. Economic margins under existing EU policy schemes (predominantly blending mandates) will remain low and comparative cost advantages will have to be used; causing a growth in production capacity in strategic locations offering diverse and cheap(er) feedstock and other input factors (e.g. labour) (Lamers 2012).

Fuel ethanol dominates global biofuel markets in terms of volumes. The primary policy schemes responsible for its growth over the past decade were tax incentives and blending mandates. Given the latter, consumption has been highest in markets with a petrol focused transport fuel matrix. The global industry has grown beyond its role model, sugarcane based Brazilian ethanol, to become dominated by US corn based ethanol production.

Table 2.2 Policy and market factors (Lamers et al. 2011)

	Stimulating domestic biofuel market	Increasing international biofuel trade
Policy		
Production related measures/policies	Investment support for local production facilities, RD&D, infrastructure projects, etc. Agricultural subsidies (e.g. EU CAP, US corn) Tax incentives in combination with import duties (e.g. VETC for fuel ethanol in US) Production mandates	Tax incentives without import duties (e.g. VETC for biodiesel in US) Differentiated export taxes (e.g. Argentina: reduced taxes for non-food products)
Consumption related measures/policies	Consumption mandates or incentives targeting domestically produced biofuels in combination with trade measures limiting biofuel imports (e.g. eligibility criteria under mandates such as undenatured ethanol in some EU MS)	Consumption mandates or incentives that do not discriminate the type or origin of the biofuel (e.g. blending mandates in the EU leading to a diversification of biodiesel feedstock)
Trade related measures/policies	Import duties/taxes Technical standards Sustainability criteria (if fulfilled by domestic production and sufficient, cost and GHG efficient biomass available; or criteria hard to fulfill by international imports)	Tariff preferences Varying tariff/duty levels stimulating alternative or triangular trade Sustainability criteria (if not sufficient, cost or GHG efficient biomass available in export destination and criteria fulfillment in exporting country is possible)
Market		
(Long-term) Market factors	Strong agricultural sector: existing infrastructure for feedstock production and processing including (strong) market players with respective know-how, networks, and associations (driving political support) Availability of cost efficient domestic feedstock Imbalanced transport fuel matrix guarantees a long-term market for investors and traders of respective biofuel substitute(s)	Agricultural export orientation Preferential climatic conditions (i.e. biomass potential) General lack of feedstock production potential in export destination (long-term) or adverse climatic conditions affecting volumes and/or prices of domestic feedstock (short-term)

(continued)

Table 2.2 (continued)

	Stimulating domestic biofuel market	Increasing international biofuel trade
Short-term market factors in regards to the EU and US	Decrease in crude oil prices significantly reduces production costs of grain and oilseed derived biofuels	Increase in crude oil price enhances the cost competitiveness of efficiently produced biofuels (esp. sugarcane derived fuel ethanol from Brazil)

CAP Common agricultural policy (EU), *MS* Member State (EU), *RD&D* Research, development and demonstration, *VETC* Volumetric excise tax credit (US)

The future of the biofuel sector will be strongly influenced by sustainability standards/requirements, focusing on land-use (conversion), conservation of highly biodiverse areas and areas with high natural carbon stock, and minimum greenhouse gas (GHG) emission reductions (see Chap. 10 of this book for more details).

References

AM. (2012). *Producers seek mass balance clampdown*. London: Argus Media.

BP. (2012). *Statistical review of world energy*. Available at: http://www.bp.com/sectionbodycopy.do?categoryId=7500&contentId=7068481

Ecofys. (2012). *Progress in renewable energy and biofuels sustainability*. Utrecht: Ecofys. Available at: http://www.ecofys.com/en/publication/progress-in-renewable-energy-and-biofuels-sustainability/

EIA. (2012). *Monthly energy review*. Washington, DC: US Energy Information Administration (EIA).

ePURE. (2012). *Rob Vierhout: Secretary-general of ePURE*. Brussels: European Renewable Ethanol.

Eurobserver. (2012). *Biofuels barometer – Various issues*. Observ'er, IJS, ECN, Eclarean, EC BREC. Available at: http://www.eurobserv-er.org

EUROSTAT. (2012). *Data explorer – nrg_1073a*. Available at: http://epp.eurostat.ec.europa.eu/portal/page/portal/eurostat/home

EUROSTAT. (2013). *Data explorer – EU27 trade since 1995 by CN8*. Available at: http://epp.eurostat.ec.europa.eu/portal/page/portal/eurostat/home

FAO. (2012). *FAO trade and markets*. Rome: Food and Agriculture Organization (FAO) of the United Nations. Available at: http://www.fao.org/economic/est/en/. Accessed Sept 2012.

Hamelinck, C., Koper, M., Berndes, G., Englund, O., Diaz-Chavez, R., Kunen, E., & Walden, D. (2011). *Biofuels baseline 2008*. Available at: http://ec.europa.eu/energy/renewables/studies/doc/biofuels/2011_biofuels_baseline_2008.pdf. Accessed 5 Feb 2012.

Kaditi, E. (2009). Bio-energy policies in a global context. *Journal of Cleaner Production, 17*, S4–S8.

Kutas, G., Lindberg, C., & Steenblik, R. (2007). *Government support for ethanol and biodiesel in the European Union*. Geneva: Global Subsidies Initiative of the International Institute for Sustainable Development. Available at: http://www.iisd.org/gsi/sites/default/files/subsidies_to_biofuels_in_the_eu_final.pdf

Lamers, P. (2012). *International biodiesel markets – development in production and trade*. Berlin: Union zur Förderung von Oel- und Proteinpflanzen (UFOP)/Ecofys. Available at: http://www.ufop.de/downloads/EV_Ecofys-UFOP_en_2012.pdf. Accessed Feb 2012.

Lamers, P., Hamelinck, C., Junginger, M., & Faaij, A. (2011). International bioenergy trade – A review of past developments in the liquid biofuels market. *Renewable and Sustainable Energy Reviews, 15*, 2655–2676.
Licht FO. (2012). Various issues. In *World ethanol & biofuels report*. Kent/Ratzeburg: Informa UK Ltd. Available at: http://www.agra-net.com/portal2/showservice.jsp?servicename=as072
Pelkmans, L., Kessels, K., & Bole, T. (2009). *Induced market disturbances related to biofuels*. Available at: http://www.elobio.eu/fileadmin/elobio/user/docs/Report_ELOBIO_subtask_2_3v6.pdf. Accessed Sept 2012.
REA. (2010). *Renewable transport fuels group meeting*. UK Renewable Energy Association. Available at: http://www.r-e-a.net/member/renewable-transport-fuels
Rosillo-Calle, F., Walter, A., & Pelkmans, L. (2009). *A global review of vegetable oils, with respect to biodiesel* (IEA bioenergy task 40). Available at: http://www.bioenergytrade.org/downloads/vegeTable 2.oilstudyfinaljune18.pdf
USCH. (2004). American jobs creation act. In *PL 108–357 (H.R. 4520)*. Washington, DC: US Congress House.
USDA. (2009). *EU27 – Biofuels annual*. The Hague: US Department of Agriculture, Foreign Agricultural Service. Available at: http://gain.fas.usda.gov/Recent%20GAIN%20Publications/General%20Report_The%20Hague_Netherlands-Germany%20EU-27_6-15-2009.pdf
USDA. (2010a). *EU27 – Biofuels annual*. The Hague: US Department of Agriculture, Foreign Agricultural Service. Available at: http://gain.fas.usda.gov/Recent%20GAIN%20Publications/Biofuels%20Annual_The%20Hague_EU-27_6-11-2010.pdf
USDA. (2010b). *India – Biofuels annual*. Delhi: US Department of Agriculture, Foreign Agricultural Service. Available at: http://gain.fas.usda.gov/Recent%20GAIN%20Publications/Biofuels%20Annual_New%20Delhi_India_6-20-2012.pdf. Accessed July 2012.
USDA. (2012a). *Global Agricultural Trade System (GATS)*. Washington, DC: US Department of Agriculture, Foreign Agricultural Service.
USDA. (2012b). *Production, Supply, and Distribution (PSD) online*. Washington, DC: US Department of Agriculture, Foreign Agricultural Service.
Walter, A., Rosillo-Calle, F., Dolzan, P., Piacente, E., & da Cunha, K. (2007). *Market evaluation: Fuel ethanol* (IEA bioenergy task 40). Available at: http://www.bioenergytrade.org/downloads/finalreportethanolmarkets.pdf
Walter, A., Rosillo-Calle, F., Dolzan, P., Piacente, E., & da Cunha, K. (2008). Perspectives on fuel ethanol consumption and trade. *Biomass and Bioenergy, 32*, 730–748.

Chapter 3
Global Woody Biomass Trade for Energy

Patrick Lamers, Didier Marchal, Jussi Heinimö, and Florian Steierer

Abstract This chapter presents global woody biomass production and trade developments and extracts energy-related trade volumes. It shows that direct, policy-influenced trade for energy has reached over 300 PJ by 2010. The majority of this volume comprises of wood pellets and wood chips aimed for consumption in the European Union (EU). Wood pellets are the largest single commodity stream and have seen a rapid production growth and trade internationalization. This is primarily due to past and expected future EU demand developments in the industrial segment, i.e. large-scale use of wood pellets in co- and mono-firing installations. Belgium, the Netherlands, the United Kingdom, and Denmark in particular are bound to increase consumption, and will remain net pellet importers. Wood pellet production has become a key diversification strategy of many forest companies and other traditional forest sectors, e.g. pulp and paper. Even energy utilities themselves are investing upstream. Wood chip trade for energy is largely limited to wood waste and small volumes of virgin wood chip (including roundwood) trade for energy in the Baltic Sea region and towards Italy. Policy-influenced fuelwood trade is also largest in Europe where it is mainly used in residential heating. Trade is predominantly regional or cross-border, and has been driven by local market price differences, winter condi-

P. Lamers (✉)
Copernicus Institute, Utrecht University, Utrecht, The Netherlands
e-mail: p.lamers@uu.nl; pa.lamers@gmail.com

D. Marchal
Wildlife and Forestry Department, Namur, Belgium
e-mail: Didier.Marchal@spw.wallonie.be

J. Heinimö
Lappeenranta University of Technology, Varkaus, Finland
e-mail: Jussi.Heinimo@lut.fi

F. Steierer
Food and Agriculture Organization of the United Nations (FAO), Rome, Italy
e-mail: wood-energy@fao.org

M. Junginger et al. (eds.), *International Bioenergy Trade: History, status & outlook on securing sustainable bioenergy supply, demand and markets*, Lecture Notes in Energy 17,
DOI 10.1007/978-94-007-6982-3_3,

tions, and regional supply shortages. Market factors and policies have both defined woody biomass trade volumes while policy changes did not have as dramatic effects on trade developments as in the liquid biofuel sector. Economic viability is the key limiting trade factor for woody biomass 'commodities'. Most exporting countries have low feedstock costs and already existing wood processing industries.

3.1 Background, Objective, and Methodology

Woody biomass is the most traditional form of energy use. To the present day, it still makes up about 50 EJ or 10 % of global primary energy supply (Edenhofer et al. 2011). Woody biomass use for heating and cooking dominates end-use and its consumption is still increasing in most developing countries (Johnson et al. 2010). Policies aiming at expanding the use of renewable energy have caused a renaissance of woody biomass use in many industrialized countries over the past decade. With this, two distinct trends have emerged: Formerly rather regional markets are increasingly integrated in global trade portfolios, and new entities, e.g. energy utilities, have become wood fuel producers (and traders). At the centre of these developments stands the European Union (EU) whose internal trade between 2000 and 2010 summed up to two-third of global solid biofuel trade (Lamers et al. 2012a).

The main objective of this chapter is to describe the past developments and the current status of global woody biomass trade for energy (Table 3.1, Fig. 3.1). Agricultural by-products such as palm kernel shells are not covered as they have so far only contributed marginal volumes to global energy related, and policy-driven

Table 3.1 Trade codes of CN/HS chapter 44 'Wood and articles of wood, wood charcoal', typically used in woody biomass for energy trade

CN/HS	CN/HS	Code definition
4401	440110	Fuelwood (logs, billets, twigs, faggots or similar forms)
	440121	Wood in chips or particles (coniferous)
	440122	Wood in chips or particles (non-coniferous)
	440130[a]	Sawdust and wood waste and scrap, whether or not agglomerated in logs, briquettes, pellets or similar forms
4402	440200	Wood charcoal (including shell or nut charcoal), whether or not agglomerated
4403	440320	Wood in the rough, whether or not stripped of bark or sapwood, or roughly squared *coniferous*
	440391	As 440320, for *oak*
	440392	As 440320, for *beech*
	440399	As 440320, for *other (poplar, eucalyptus, birch)*

[a]Replaced by codes HS 440131 Wood pellets and HS 440139 Other as of 2012. The respective EU codes are CN 44013100 Wood pellets, CN 44013910 Sawdust of wood whether or not agglomerated in logs, briquettes, or similar forms (excl. pellets), and CN 44013990 Wood waste whether or not agglomerated in logs, briquettes, or similar forms (excl. sawdust and pellets)

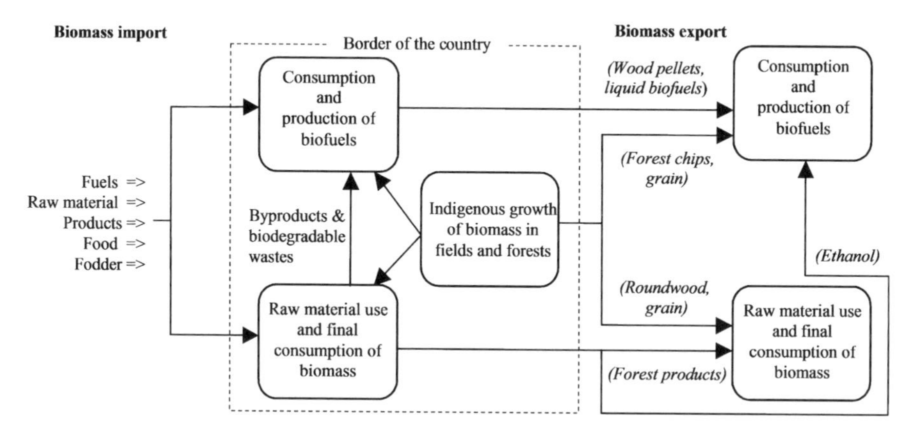

Fig. 3.1 Simplified illustration of biomass trade flows for energy (Source: Heinimö 2008)

trade (Lamers et al. 2012a). Apart from providing quantitative overviews, it also elaborates on the market drivers, and provides a methodological assessment of the respective influencing factors. The analysis is limited to direct trade of commodities for modern bioenergy use in markets where bioenergy support policies are in place. Indirect trade, i.e. volumes not directly related to energy usage, such as wood chips of which a fraction ends up as black liquor and thus energy, are not examined in detail. The chapter however includes an exemplary assessment for the period 2004–2011. Also, woody biomass trade in markets where no bioenergy support policies are in place, e.g. fuelwood or charcoal use across sub-Saharan Africa, is not included since an integral part of the IEA Bioenergy Task 40 work has been the evaluation of policy influences on bioenergy market and trade flow developments.

World production data is derived from (FAOSTAT 2013). Trade data was collected via UN 2013 and EUROSTAT 2013 statistics. Trade codes only refer to the physical appearance of commodities and not their final end-use so that energy related trade streams have remained rather informal. Hence, official data sources needed to be sidelined with anecdotal evidence, e.g. from previous IEA Bioenergy Task 40 work (Bradley et al. 2009; Cocchi et al. 2011; Lamers et al. 2012b; Goh et al. 2013), and scientific methodologies to account for energy related trade only (Heinimö 2008; Heinimö and Junginger 2009; Lamers et al. 2012a) (Fig. 3.1).

The next section provides a global overview of woody biomass production and trade. Global energy related trade developments are inherently part of these trade streams but need to be extracted. This is done in the following section, which also provides exemplary calculations of indirect trade flows. Afterwards, the currently largest trade streams (wood chips and wood pellets) are presented in detail and with specific link to the currently most important market and trade centre: Europe. Wherever applicable, European developments are put into a global perspective. Second to last, a separate section shows which policy and market factors have affected these past developments. The chapter closes with concluding remarks and an outlook on possible future trade developments.

3.2 Global Production and Trade Developments

The total production and relative trade shares per annual production volume of the selected commodities (Table 3.1) are given in Fig. 3.2 (see e.g. FAO 2012 for additional information and commodities). Roundwood is the largest absolute trade stream, followed by fuelwood, wood chips, charcoal, and sawdust and wood waste (Fig. 3.2 top). Traded production shares of the commodities show a heterogeneous picture: While charcoal and fuelwood shares remained stable, wood chip trade shares declined, and sawdust and roundwood shares showed upward trends (Fig. 3.2 bottom).

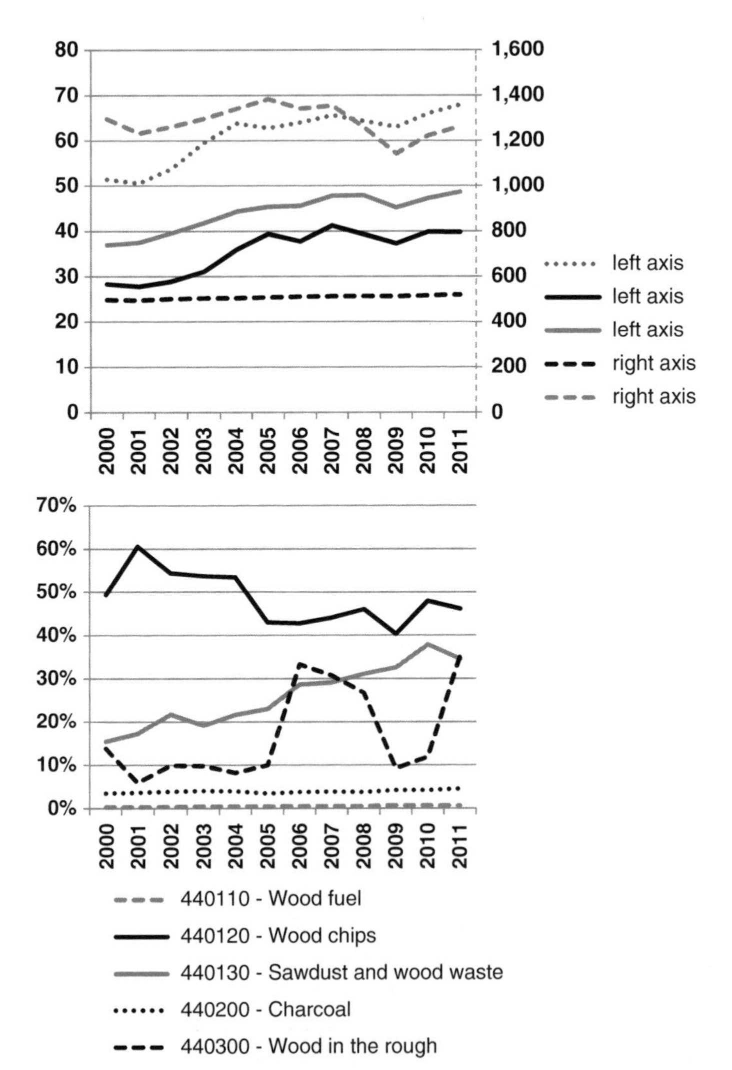

Fig. 3.2 Total global production (*top*) [in Mtonnes] and share of annual production traded (*bottom*) of selected woody biomass commodities (Data: FAOSTAT 2013; UN 2013)

Until 2005, between 10 and 15 % of global industrial roundwood (HS 4403) production was traded annually. Later, trade volumes rose significantly with over 30 % of global production being traded in 2006–2008 (Fig. 3.2). The drop in 2009 is attributed to the introduction of export duties on roundwood in Russia and the global financial crisis. By 2011, trade volumes have risen again to 2006-levels, i.e. 446 Mtonnes (Fig. 3.2), representing 35 % of annual production (Fig. 3.2). This shift in trade shows that roundwood had typically been processed locally; as core forestry nations have also been home to the world's major wood processing industries. More and more however is processed elsewhere, particularly in Asia. The largest absolute production increases have taken place in Brazil, Russia and Indonesia. The largest declines were noted in the USA and Canada. By 2010, five countries produced half of the total world production: USA (20 %), Russia (9 %), Canada (8 %), Brazil (8 %), and China (7 %). While the majority of global roundwood production and trade is not connected to bioenergy, there is a large amount of indirect trade in the form of wood processing residues (Heinimö 2008).

Wood chips (HS 440121, HS 440122) represent the second largest absolute single trade stream (Fig. 3.2), and mainly consists of high quality chips for pulp and paper production. Global wood chip production has grown to 68 Mtonnes by 2011, while the traded annual production volume has declined over time (Fig. 3.2). This may be connected to the general shift of production to economies with low labour costs, predominantly in the Southern hemisphere. The largest production increases over the past years have taken place in China, South Africa, and Brazil. The strongest production decline was noted in the US. By 2010, the key producing nations included Canada (31 %), China (8 %), Australia (7 %), and Sweden (7 %); all major pulp and paper producers. Australia and Canada are also key wood chip exporters – predominantly to Japan and other Asian countries. Trade in wood chips for energy (virgin and/or tertiary waste) is practically limited to Europe, Turkey, and Japan (Lamers et al. 2012a).

Global production of wood residues (as previously covered under HS 440130) has grown from 28 Mtonnes in 2001 to 40 Mtonnes in 2011 (Fig. 3.2). Over the same time, the traded annual production share has increased from 15 to 35 %. The largest part of this volume is made up of harvesting (tops and branches) and processing (sawdust) residues in the form of wood pellets. Wood pellets have become the largest single energy-policy related trade stream. As of 2012, they are tracked under a specific trade code (HS 440131). The remaining fraction, previously listed as 'waste wood and scrap' has now to be filed under HS 440139.

Large scale international shipments of recycled wood for energy purposes are still rare but have been known to occur (Lamers et al. 2012b). Up to now however, the larger share of wood waste is generally landfilled, combusted locally or traded short distances. The key region for international wood waste trade is currently Europe; primarily due to its differences in legal and bioenergy policy frameworks across the individual Member States (Lamers et al. 2012a). Wood waste is generally not chipped but rather crushed to minimize transportation costs.

Fuelwood (HS 440110) and charcoal (HS 440200) represent the lowest annual trade shares. Both can be considered local products, with less than 1 % (fuelwood) or 5 % (charcoal) of their annual production being traded respectively. These numbers however are based on official statistics and are unlikely to include informal cross-border trade, which may be significant given the high share of either fuel in traditional heating and cooking. The same statistics indicate that fuelwood production remained relatively stable over the past decade, while charcoal production continuously increased.

The higher international trade share for charcoal compared to fuelwood (Fig. 3.2) can be explained by the additional uses of charcoal. Apart from heating and cooking (including barbeque in industrial countries), charcoal is applied in the chemical (as active coal) and in the iron and steel industry (as a reducing agent and energy source). The largest producer between 2000 and 2010 was Brazil (13 %), where most charcoal is used in pig iron production. International trade with charcoal has been dominated by Germany (10 %), Japan (9 %), and South Korea (8 %) in terms of imports (Lamers et al. 2012a). Poland is the largest source for charcoal trade to Germany. Japan and South Korea tend to source their charcoal mainly from Indonesia, Malaysia, and China. Total world exports have been led by Somalia over the past 4 years (Economist 2012; FAOSTAT 2013). International charcoal trade generally takes place in bagged form. Up to now there is no direct and large scale trade for modern energy conversion, and current trade for energy purposes is limited to heating, cooking and barbeque.

Fuelwood use for heat generation in high performance boilers and stoves has been heavily driven across the EU over the last years. Its share in global trade increased from 50 % (2000–2004) to over 80 % (2007–2011). Most of this trade takes place cross-border: short- or mid-range in bagged form, conglomerated in nets, or stacked on pallets. Recorded trade streams outside Europe are between South Africa and its neighbouring countries (Swaziland and Namibia), Canada and the USA, and across South East Asia. By 2010, half of the total fuelwood production was centred in India (17 %), China (10 %), Brazil (8 %), Ethiopia (5 %), Congo (4 %), Nigeria (3 %), and Indonesia (3 %) combined.

3.3 Total Energy Related, Policy-Driven International Trade

When defining the amount of policy-influenced woody biomass trade for energy, absolute global trade streams for specific commodities (Fig. 3.2) have to be broken down into their different end-use fractions. Officially reported volumes cover energy related and other streams, e.g. for material purposes in the case of wood chips and roundwood. In addition, one needs to account for potential cross-trading such as re-exports or wholesale activities to avoid double-counting.

Official trade data (e.g. EUROSTAT 2013; FAOSTAT 2013; UN 2013) can thus be seen as a theoretical upper limit of possible energy-related trade. To obtain solely bioenergy related production and trade streams, it is yet indispensable to rely on anecdotal evidence, such as conference presentations, speeches, and interviews of internationally recognized experts from private market parties, industry, academia, or else.

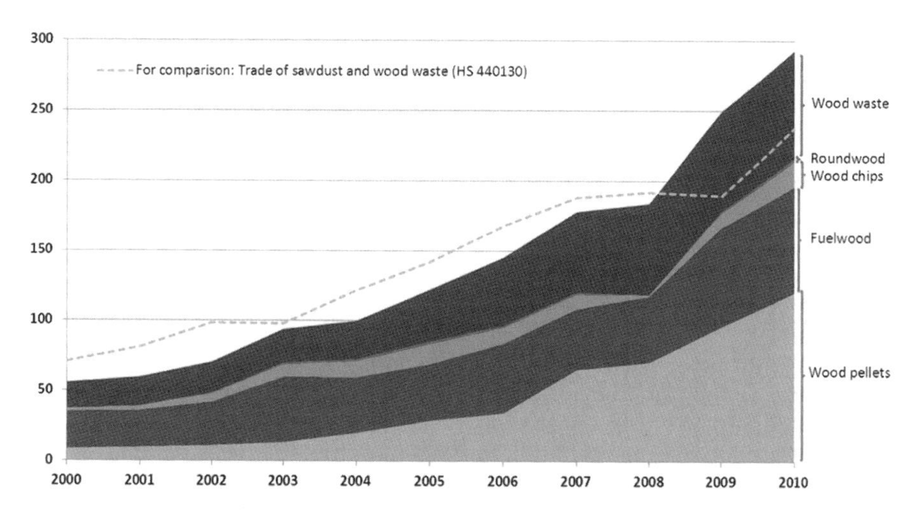

Fig. 3.3 Estimated global net solid biofuel trade [in PJ] (Data: Lamers et al. 2012a)

3.3.1 Net Trade for Energy

The methodology for global net wood pellet trade builds on the central observation (given past developments) that the most lucrative markets from a producer and trader perspective lie in the EU, and to some extent also in the US, Japan, and South Korea due to the policy influenced local market value for woody biomass (Lamers et al. 2012a).

There is practically no risk of double-counting since the markets are yet still separated – apart from the EU-internal distribution of overseas imports. Large-scale shipments also often occur directly from the producer to the end-user. Indirect trade of raw material, e.g. roundwood for wood pellet production has been neglected as the main pellet producing nations are also key roundwood suppliers and do not depend on imports. Furthermore, waste wood and direct roundwood to pellet conversion were only at their initial industry stage in 2009/2010 and the respective facilities all own local forestry plantations.

Given this framework, net woody biomass trade volumes for energy grew sixfold from 56.5 PJ (3.5 Mtonnes) to 300 PJ (18 Mtonnes) between 2000 and 2010 (Fig. 3.3) (Lamers et al. 2012a). Over this period, wood pellets grew strongest and became the dominant commodity on international markets, whereas trade with wood waste, roundwood, and wood chips for energy remained much smaller and practically limited to Europe.

Until 2002, policy-driven woody biomass for energy trade development was largely motivated by legal and technological differences for wood waste combustion in the EU. Trade for residential heating application dominated, and wood pellet trade was limited to intra-EU and intra-North American trade. After 2003, when policy schemes for the promotion of renewable energy and more specifically bioenergy derived electricity production emerged across the EU, trade with wood waste, chips, pellets, and residues started to grow. Incentives for the installation of heat stoves across EU Member States in particular benefited pellet trade whereas trade in wood chips and fuelwood only grew marginally. Regional trade fluctuations mostly adhere to winter conditions and local availability. Post 2005, we see that extra-EU production costs and EU renewable energy support drives EU pellet imports for medium and large-scale power production. Also, there is a growing industry trend towards pellet usage, i.e. away from wood chip combustion. Wood chip trade remains rather regional. This trend continues and is fuelled by rising oil prices until the financial crisis late 2008. In recent years, increases in global policy-influenced woody biomass for energy trade are attributed in principal to a growing US-EU pellet trade, further increases in EU pellet production for intra-EU trade, an oversupply of roundwood from Russia partly traded for energy in the form of wood pellets, wood chips, and fuelwood, and the increase of policy support and trade of pellets across Asia.

3.3.2 Indirect Trade of Raw Materials for Energy

Net trade estimations as shown in Fig. 3.3 do not incorporate indirect trade, i.e. volumes not directly related to energy usage, such as wood chips of which a fraction ends up as black liquor and thus energy. Nevertheless, indirect trade can sum up to substantial amounts.

The wood processing industry procures wood primarily as a raw material. In many cases, wood is imported from other countries. For example, Finland imports large amounts of raw wood (logs, pulp wood, and chips) from, e.g. Russia. In the manufacturing processes of the primary products, a significant amount of the raw wood ends up in energy production or is converted into by-products for energy generation. Biofuel purchase and use of this kind is referred to indirect import of biofuels, and the corresponding export is called indirect export of biofuels. The previously mentioned wood streams jointly constitute indirect trade of biofuels.

On average, 40–60 % of the roundwood can be converted into wood products in the forest industry. The remaining share ends up as a by-product, such as black liquor, bark, sawdust, and chips, with no material use within the specific industry. Using trade volumes for industrial roundwood and wood chips and particles from FAOSTAT 2013, we provide a rough estimate about the potential volume of indirect trade of wood based biofuels (Fig. 3.4). For both roundwood and wood chips, we assume that 45 % is the total trade volume is converted to energy with a calorific value of 9.4 GJ/tonne. Uncertainties in this calculation include, e.g. the conversion efficiency for raw wood, which varies between the production processes of different products, and the level of technology applied and the integration of the production processes which affect conversion efficiency. For a more detailed description of the methodology, we refer to Heinimö 2008 and UNECE/FAO 2010.

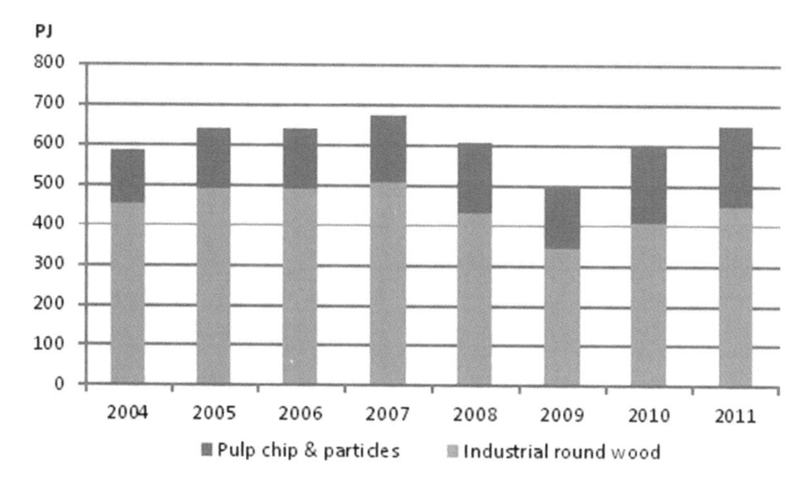

Fig. 3.4 Estimated volume of indirect trade of biomass for energy [PJ]

3.4 Wood Chip and Wood Pellet Market Developments

Industrial roundwood dominates absolute international woody biomass trade volumes. The vast majority of this trade flow however is for non-energy related purposes. Wood fuel and charcoal are traditional energy carriers and typically not traded over long distances, but rather regional or cross-border. Charcoal trade is not considered policy-driven and therefore not further examined. Wood fuel production and consumption increases across the EU have been regarded as policy-influenced, international wood fuel trade however has rather been driven by supply fluctuations and extreme weather conditions, e.g. Swedish imports from Latvia and Russia during harsh winter conditions (Lamers et al. 2012a).

Wood chips, in raw form or as wood waste, and wood pellets are currently the largest absolute energy-related global trade streams (Fig. 3.4) The wood chip trade initially shown in Fig. 3.2 is largely destined for pulp and paper production, with some trade for other uses such as fibre and particle boards. It is estimated that less than 10 % of annually reported trade volumes are energy-related (Lamers et al. 2012b). These cover wood chips made of roundwood, residues, and waste wood. The latter would have been covered under the former HS code 440130. It is noteworthy that this is the only commodity of the selection in Table 3.1 whose production and trade share volumes have continuously grown over the past decade (Fig. 3.2). The reason for this lies in the second component of the trade code: sawdust in the form of pellets.

3.4.1 Wood Chips

The key distinction between wood chips is their source material. High quality chips derived from roundwood are a valuable component in pulp and paper production. Wood chips for energy purposes can either be derived from recovered/waste wood or virgin wood.[1] The latter are typically derived from harvesting residues such as branches, tops, thinnings, or other inferior wood not suitable for material or pulp and paper production, or other processing residues. This implies that bioenergy related trade streams may have fallen under various trade codes including coniferous (HS 440121) or non-coniferous wood chips (HS 440122), or sawdust and waste wood (HS 440130). Technically, wood chips may also be transported as fuelwood (HS 440110) or roundwood (HS 4403) prior to chipping and combustion.

Across Europe, there are two distinct virgin wood chip markets for energy and thus major trade flows. The first encompasses the Baltic Sea bordering states, where Sweden and Denmark (and to some extent also Germany) have been leading importers over the past decade, sourcing largely from Russia and the Baltic states (Junginger et al. 2010). The second market lies in southern Europe, primarily driven by Italian facilities sourcing from neighbouring countries, the Balkan in particular (Fig. 3.5).

Virgin wood chips for the EU residential market are primarily sourced locally. International wood chip trade is exclusively driven by the industrial sector, where chips are combusted in dedicated or converted co- and/or mono-firing installations (primarily fluidized-bed). Respective trade takes place in the form of (virgin) wood chips, crushed (waste) wood, or as roundwood which is chipped at the plant (Lamers et al. 2012a, b). Official statistics indicate that wood waste volumes dominate the EU-related trade.

Between 2000 and 2005, (virgin) wood chips were transported for energy from the US to Europe (primarily Italy); with annual volumes of up to 200 ktonnes (Flynn, 2012, May, Director International Timber at RISI, USA: Global wood chip trade patterns for energy, personal communication). When it became apparent that these streams were in violation of the EU requirements for phytosanitary measures (see Lamers et al. 2012b for details), the trade was stopped. The EU still requires phytosanitary measures for softwood chips from North America. The restrictions have practically eliminated the largest of the softwood chip trade (utilizing Southern Yellow Pine) for both energy and pulp and paper production to Europe (Guizot 2010). As a result, softwood chip streams from North America to countries with less import restrictions have grown, in particular to Turkey and China.

In 2010, there was a substantial increase in demand for wood chips in China. The nation has evolved from being a net exporter of chips 5 years ago, to being a major chip consumer, having quadrupled imports in just 2 years. The country now imports over 28 % of all chips traded in the Pacific Rim and is the world's second largest importer of wood chips after Japan. Trade of wood chips is still the highest in the Pacific Rim, accounting for almost 60 % of the total global trade and over 95 % of water-born trade. So far, little is known about exact volumes entering China for energy purposes. Naturally, Chinese imports could reach very large dimensions in

[1] The EU wood chip quality standard EN 14961–4 defines four classes (FOREST 2011). Class A1 and A2 represent wood chips from virgin wood or chemically untreated wood residues with different ash and moisture contents. Class B1 and B2 extend the source of biomass to chemically treated industrial wood by-products and residues and used wood.

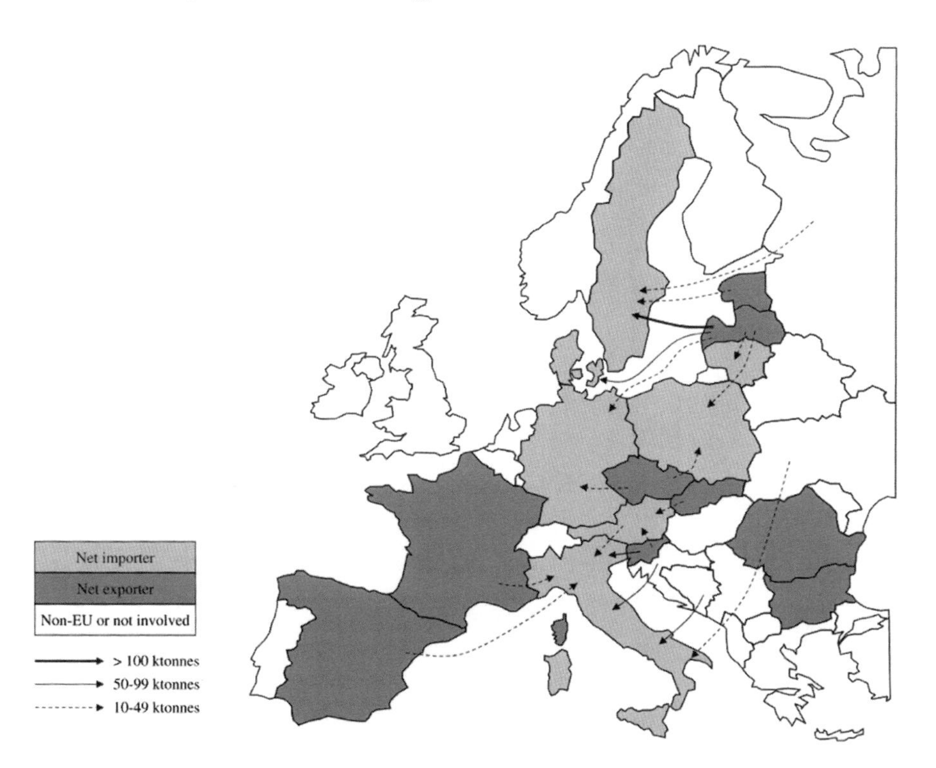

Fig. 3.5 Dimensions of virgin wood chip and roundwood trade for energy (Data: Lamers et al. 2012a, b; EUROSTAT 2013. Note: Trade streams towards Denmark, Germany and Sweden are also indicators for roundwood trade volumes and routes.)

the near future. Apart from North America, trade flows to China originate primarily in Oceania and Asia (including Australia, Vietnam, and Russia).

Japan has also previously sourced (virgin) wood chips from Canada. In 2011, 300 ktonnes wood chips were imported for energy combustion (Goto et al. 2012). Previously it was suggested that wood chips for energy production would be solely derived from domestic demolition wood and that Japanese wood chip imports cover pulp chips exclusively (Goto et al. 2012; Lamers et al. 2012a).

Norway has previously imported mostly hardwood chips from Canada, Brazil and Africa for pellet production (Willumsen 2010). Trade volumes reached around 330 ktonnes per year by 2011 (Willumsen 2010) but came to a halt by 2012 (Flynn, 2012, May, Director International Timber at RISI, USA: Global wood chip trade patterns for energy, personal communication).

For a period of 2–3 years, some of the larger international trade flows to the EU for energy purposes were rubberwood chips from Liberia destined for co- and mono-firing installations of energy utility Vattenfall in Germany and Sweden. The sourcing strategy of the utility has since changed, and, according to official statistics, Liberian trade flows to Europe ceased in 2012.

Generally, wood chips for energy purposes have been transported over shorter distances than wood pellets. Apart from vessel size restrictions in the Baltic Sea region, this is primarily due to the ratio between moisture content, heating value, and bulk density (Alakangas et al. 2007; Junginger et al. 2010).

European combustion facilities have been known to not only import virgin wood chips, i.e. previously unused, woody biomass excluding tertiary residues, but also roundwood for making wood chips. This allows facilities to control chip sizes and quality. There are also storage benefits of roundwood (moisture, heating value). In the Baltic Sea region, wood chip and roundwood trade are closely interwoven. They are often traded on the same vessel (Vikinge 2011) and destined for the same conversion facilities. Winter conditions in Baltic Sea harbours have in the past led to increasing imports from Southern Europe. So far no overseas EU imports of roundwood for energy purposes are known however.

Today, waste wood is typically (co-) combusted across the EU. Differences in renewable energy policy support schemes have driven a strong EU-internal trade of crushed or chipped waste wood (i.e. tertiary wood chips). Historically, Sweden was among the first states to attract large amounts of wood waste. Today, trade to other Member States is far larger. Top importing nations include Germany, Italy, and Belgium. The major exporters are the Netherlands and the UK. The relatively balanced import–export relation of Belgium and Germany is largely related to national policy schemes which favour different streams of waste wood. The German renewable electricity feed-in scheme e.g. has provided strong incentives for the combustion of clean (non-treated) waste wood. Whereas, in the past, more contaminated waste wood had e.g. attracted higher subsidies in the Netherlands (Faber et al. 2006).

3.4.2 *Wood Pellets*

By 2010, around 60 % of global wood pellet production was concentrated in the EU (Fig. 3.6). Since 2000, EU production, demand, and imports have increased more than tenfold (Lamers et al. 2012a). This trend is clearly policy-influenced. Production and trade patterns have developed in accordance with the respective consumer markets in the individual Member States. Until 2010, most pellets were combusted in residential heating (dominated by Italy, Germany, and Austria), followed by district heating (Sweden and Denmark), and large-scale power production (concentrated in Belgium, Denmark, the Netherlands, and the UK) (Hiegl and Janssen 2009; Sikkema et al. 2011).

We can distinguish between two different pellet types across these markets. Previously, this distinction has been coined high (residential) vs. low (industrial) quality or white (residential) vs. brown (industrial) pellets. These attributes are however relative and not always applicable to the respective trade flows. We suggest differentiating instead merely between end-use, i.e. residential or industrial

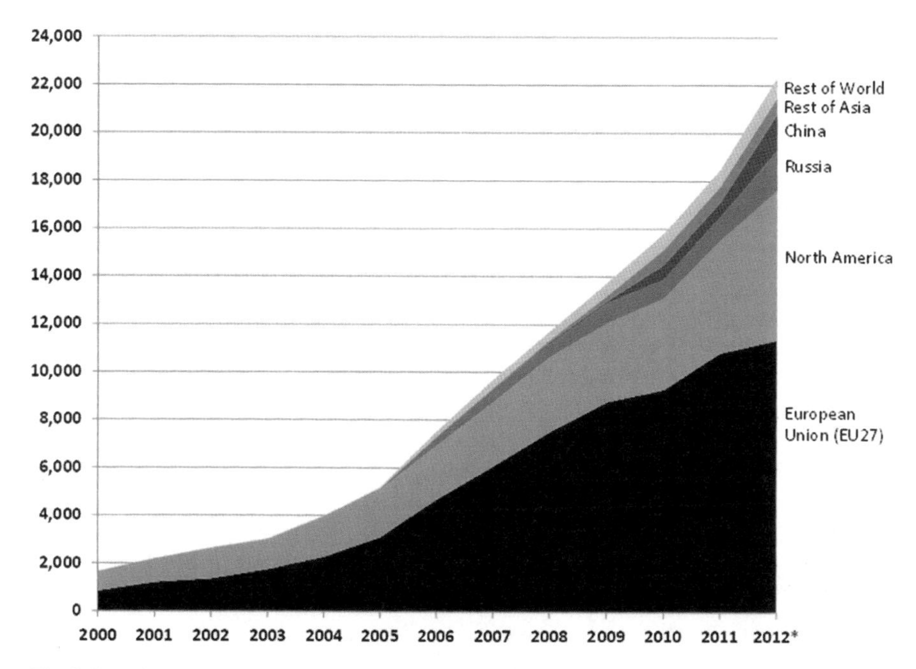

Fig. 3.6 Estimated global wood pellet production [in ktonnes] (Data: Cocchi et al. 2011; Lamers et al. 2012a; REN21 2012; EUROSTAT 2013; Goh et al. 2013)

pellets.[2] Residential pellets generally need to fulfill higher restrictions regarding ash content, ash melting point (slacking), and water content (fouling). Requirements for industrial pellets though have also increased and there are even ambitions to create a uniform certificate (ENplus) for pellets used or traded for the EU heat and power markets, applying to intra-EU trade and extra-EU imports.[3]

Residential pellets have been exclusively sawdust derived. Industrial pellets have also been largely sawdust derived, while the shares of harvesting residues including bark, tops and branches of merchantable trees (for timber) plus whole non-merchantable trees, and pulpwood-quality roundwood have increased.

A key difference between the markets so far has been the logistical aspect and the sourcing range. Residential pellets have largely been supplied in bulk or bagged form via regional retailers and wholesalers. Very little global trade has taken place. Industrial pellets however have been transported globally in large bulk quantities and directly sold and shipped to the power plant/consumer. Very little retailer/wholesaler activity takes place in this segment. It is expected that future trade in the residential sector will also encompass global trade in large bulk quantities. First shipments between Canada and Denmark have already taken place.

While being relatively self-sufficient in the residential pellet market segment, the EU has become heavily import-dependent in the industrial pellet market. This is largely

[2] The EU wood pellet quality standard EN14961-2 differentiates between categories A1, A2, B.

[3] See http://www.enplus-pellets.eu/pellcert/ [January 2013].

Table 3.2 The 15 largest operating pellet mills by 2012

Company	Location	Predominant feedstock	Capacity [tonnes/year]
Vyborgskaya Cellose	Russia	Unmerchantable timber	900,000
Georgia Biomass	GA, USA	Plantation roundwood	750,000
Green Circle (JCE Group)	FL, USA	Plantation roundwood	500,000
Biowood	Averøy, Norway	Wood chips (imported)	450,000
Pinnacle Pellet Inc	BC, Canada	Sawdust, pine-beetle wood	400,000
Enviva, Hertford	NC, USA	Unknown	350,000
Pacific BioEnergy	BC, Canada	Sawdust, pine-beetle wood	350,000
German Pellets	Wismar, Germany	Sawdust	256,000
German Pellets	Herbrechtingen, Germany	Sawdust	256,000
Arkaim	Khabarovsk, Russia	Processing residues	250,000
Plantation Energy	Albany, Australia	Eucalyptus residues	250,000 [closed]
Pinnacle Pellet Meadowbank	BC, Canada	Sawdust, pine-beetle wood	220,000
Ankit Pellets & Briquettes	Bengaluru, India	Unknown	200,000
Houston Pellet Inc	BC, Canada	Sawdust, pine-beetle wood	180,000
Graanul Invest Incukalns	Incukalns, Latvia	Sawdust	180,000

due to the increase in demand (predominantly in the Netherlands and the UK), but also linked to a limited mobilisation and production (within the EU and its border countries), competitive to overseas pellet export prices. Exports from Fennoscandia via the Baltic Sea are limited by ship size and frequently also by ice forming in harbours in winter – the peak demand season for district heating installations, e.g. in Denmark.

In addition, to the increasing number of pellet producers looking to supply into a growing European market, several European energy utilities themselves have started (or are aiming) to expand their activities upstream to secure their supplies of industrial pellets. The strongest absolute production increases in recent years took place in North America, more specifically in the South-East USA (Table 3.2). While also having a strong local consumption, much of the additional US capacity installed since 2010 is aimed at producing industrial pellets for export to the EU (Goh et al. 2013). In 2009, about 1.7 Mtonnes were imported from outside the EU. By 2012 this volume had risen to 4.6 Mtonnes (Fig. 3.7). By 2020, we expect EU wood pellet imports to be in the range of 15–20 Mtonnes (see also Cocchi et al. 2011; Goh et al. 2013). While the majority of this volume will be industrial pellets, wood pellet imports for the domestic market are also expected to increase.

Sourcing of wood pellets from many different world regions creates a challenge for companies and policy makers to ensure that the biomass used is from sustainable sources. The vast majority of wood pellets are still derived from either harvesting or processing residues, largely tops and branches from lumber harvest and sawdust and chips from timber sawmills. The expansion of the sector however, as new investments have shown, will inevitably lead to the use of lower quality roundwood

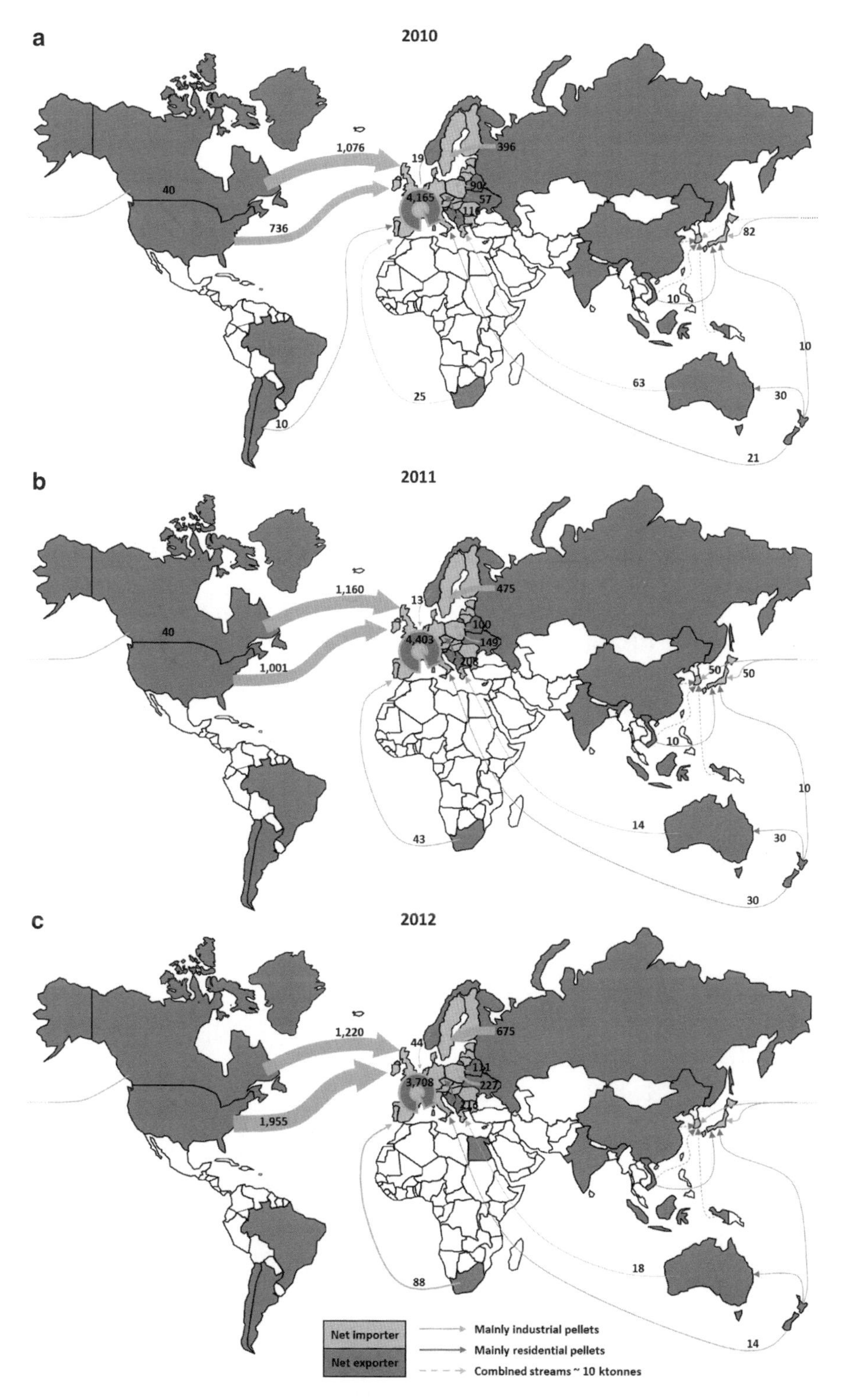

Fig. 3.7 Global wood pellet trade streams in (**a**) 2010, (**b**) 2011, (**c**) 2012 [>10 ktonnes] (Data: Cocchi et al. 2011; Lamers et al. 2012a; EUROSTAT 2013)

fractions (e.g. pulpwood). This feedstock switch may involve new sustainability risks, with potential impacts on forest carbon stocks, biodiversity, and soil productivity being the most pressing (Lamers et al. 2013).

3.4.3 Country Examples: Wood Pellet Markets

Based on Cocchi et al. 2011; Goh et al. 2013 summarized the characteristics, policies/regulatory framework conditions, and trends for the most important wood pellet markets. They distinguished them by consumption pattern and trade profile; Europe and Asia being the key import regions and North America dominating the export focused market side. Table 3.3 provides one country example per specific market category. The European market is differentiated by dominating end-use within the respective Member State.

3.5 Policies and Market Factors Shaping International Woody Biomass for Energy Trade

The different trade pattern developments of the woody biomass commodities imply foremost that they were exposed to different demand, supply, or trade stimuli and show different biofuel or market characteristics (Table 3.4). The dominant producing and exporting nations are clearly those with long-standing, export-oriented forestry, wood processing, and/or pulp and paper industries. The availability of excess residues, the possibility to use existing infrastructure (for processing and transport), and interconnected know-how have turned out to be key drivers and success factors. This is particularly true for regions where domestic policies have triggered national bioenergy markets, as e.g. in Scandinavia or Austria. Countries with little policy support or interest in bioenergy, due e.g. to the abundance of fossil fuel resources (e.g. Russia) show that the existence of aforementioned key factors under absence of strong national interests is not necessarily a guarantee for success. Slow and inconsistent developments in young markets with a high theoretical biomass potential (see e.g. Smeets et al. 2007) further exemplify the need for local experience and continuous domestic interest for market off-taking.

The key defining factor for international solid biofuel trade is economic viability (Lamers et al. 2012a). On the supply side, it is constrained by production and transport costs; in turn mostly influenced by feedstock, vehicle costs, and biofuel characteristics (Olsson et al. 2010). The heating value, correlated to moisture and ash content (Kaltschmitt and Hartmann 2001), bulk density, (homogeneous) form and chemical composition define the monetary value as a biofuel. These factors determine whether a commodity is worth transporting over long distances (via ship), relatively more expensive (short) transport modes (e.g. via truck), or whether it requires further processing (e.g. drying, pelletizing, torrefying; see also Uslu et al. 2008). Low heating

Table 3.3 Country examples for different wood pellet net demand and net supply markets (Goh et al. 2013)

Market characteristics	Policies/Regulatory framework	Market trend
Europe: Residential and district heating: Germany		
No co-firing of wood pellets in power plants – rely on other renewable electricity such as solar, wind and other biomass power	Market incentive program (MAP) – investment subsidy Renewable Energies Heat Act (EEWärmeG) – building regulation	Increased use of wood pellets for residential heating Depletion/Freezing of MAP budget caused uncertainties among the investors leading to a smaller number of pellet heating systems installed in 2010 Reinforced utilization of pellets on the small-scale market The pellets produced for power generation are entirely exported
Europe: Power plants driven market: the UK		
Transparent monitoring of biomass energy use and sustainability certification by Ofgem	Renewable Obligation Electricity Market Reform Feed in Tariff	Increasing use of wood pellets, largely in power plants
Europe: Mixed market: Sweden		
Opportunities: High oil prices, increasing electricity costs, and heavy taxation on fossil fuels Barriers: High raw material prices and intense competition	Electricity certificate system combined with renewable obligations and exemptions from CO2 taxes Indirect effect: Heavy fossil fuels tax	Use of wood pellets in private households has increased by a factor of 20 over a 13 year period Raw material shortage
Other import oriented markets: South Korea		
Large coal power plants	Renewable portfolio standards for power companies (by 2012) Korea Forest Service subsidize the purchase of domestic pellet boilers by 60–70 %	Possibility to co-fire wood pellets with coal Expected that at least 60 % of renewable energy will be from pellets, amounts to 2.25 Mt in 2012 Induce pellets production in Indonesia, Myanmar and New Zealand About 13,600 boilers were installed since 2008
Export oriented markets: the USA		
Many underutilized sources of biomass – mill residues and crop residues	Federal level: Renewable Energy Production Incentive (REPI)	Over 80 % of pellets produced in the US were used domestically; of the remaining, were exported to Europe Production declined when the sawmilling sector retrenched in the 2008–2009 recession

(continued)

Table 3.3 (continued)

Market characteristics	Policies/Regulatory framework	Market trend
Reliance on sawmill residues Increases in the cost of fossil energy Demand and also investment from Europe	State biomass economic drivers	New mills to process chipped roundwood – independence from the sawmill industry has allowed a focus on export Regulations will likely drive existing coal power plants to co-fire with biomass, which will create an increasing market for biomass pellets

Table 3.4 Trade influencing market & policy factors (Lamers et al. 2012a)

	Stimulating (local) solid biofuel production/use	Stimulating international solid biofuel trade
Market		
Market characteristics	Availability of excess residues from existing forestry, pulp and paper, or wood processing industries; also allowing the use of the respective infrastructure, know-how, and political influence Preferential climatic conditions (i.e. potential) Existing businesses with facilities allowing co-/mono-firing; especially fluidized bed technology due to feedstock flexibility	Existing export orientation of the forestry or wood processing industry: big scale bulk infrastructure (railways, harbors), handling equipment (chippers, cranes, terminals etc.), export market/trade know-how High local electricity and heat prices increasing the economic viability for biofuel imports Availability of low cost domestic fossil fuels (e.g. in Russia, North America) allowing/stimulating exports of low cost domestic biofuels Limited large-scale, low cost, domestic feedstock production potential
Solid biofuel characteristics	Local (short-distance) use is typical for biofuels which are either unrefined, cannot be transported in bulk (fuelwood), have a high moisture content, low monetary and/or low heating value (e.g. forestry slash, bark chips) Small margin between supply costs (production and transport) and prices in consumer markets	Refined, homogeneous biofuels with high heating and/or monetary value (e.g. pellets), bulk density, flow-ability (reducing handling costs); low moisture and ash content Large margin between supply costs (production and transport) and prices in consumer markets Similar combustion characteristics to coal increasing the attractiveness for co-firing Flexible end-use (combustion technology and scale)

(continued)

Table 3.4 (continued)

	Stimulating (local) solid biofuel production/use	Stimulating international solid biofuel trade
Policy		
Supply related	Incentives to increase the residue use in the forestry and/or agricultural sector, or the planting of dedicated cellulose crops via investment support, direct subsidies, low-interest loans, grants, or infrastructure projects Mobilization of small forest owners (see e.g. EC 2010 on good practice guidance on the sustainable mobilization of wood in Europe)	Overproduction due to lack of local demand, overstimulation and/or highly competitive production prices compared to other international sources incentivizing exports Differing legal requirements for waste wood combustion between neighbouring countries
Demand related	Renewable electricity and/or heat targets enforced via regulatory or fiscal policies Emission standards Ban on landfilling wood waste Investment support via low-interest loans, grants, or subsidies for equipment	Linking domestic policies with eligibility criteria i.e. limiting the combustion to certain biofuel streams thus increasing respective imports and triggering exports of non-eligible material; the same is true under limited national potential
Trade related	Commodity specific export duties Technical standards e.g. in the form of phytosanitary measures for imports Hypothetically also sustainability criteria (if fulfilled by domestic production and sufficient, cost and GHG efficient biomass available; or criteria hard to fulfil by international imports)	Avoidance of export duties by transforming the respective commodity Technical standards in the form of globally accepted quality standard (e.g. ENplus for wood pellets) Hypothetically also sustainability criteria (if not sufficient, cost or GHG efficient biomass available in export destination and criteria fulfilment in exporting country is possible)

value products (e.g. forestry residues) with a relatively low monetary value are usually used locally or transported cross-border. Refined, homogeneous biofuels with a high heating (low moisture and ash content) and monetary value (e.g. bagged wood pellets for residential heating) are traded globally. A high bulk density is preferable for long or expensive transport and is clearly influenced by processing (e.g. condensing into wood pellets).

On the demand side, the defining economic factors for trade are the margins achievable in the respective end-consumer markets. They are typically influenced by local policies, exchange rates (Olsson et al. 2010), and application scales (Lamers et al. 2012a). Economies of scale allow the trade of low value commodities, in particular those with combustion characteristics similar to those of coal and/or which may be used in different technical installations.

There are multiple design options for renewable energy support schemes (see Mitchell et al. 2011 for a review). They are generally able to influence markets on the supply and demand side, but also trade via tariff policies and regulations. Tariffs have largely been irrelevant as a most-favoured nation (MFN) tariff of 0 % applies between the members of the World Trade Organization (WTO). Trade regulations however did have impacts on trade flows as the EU requirement for phytosanitary measures for wood chips from North America has shown.

Supply side policies predominantly aim at reducing production costs, i.e. costs for feedstock and processing (Alakangas and Keränen 2011). Indirect stimuli via research and development support (e.g. forestry management practices) or direct subsidies for equipment, the collection of unused residues, or planting of dedicated cellulose crops are other examples. As supply side policies have primarily stimulated domestic markets, they yet had marginal effects on trade.

This is different to demand side policies which have significantly stimulated international trade (Lamers et al. 2012a). EU developments illustrate the varying policy effects nicely (see Ragwitz et al. 2007; Held et al. 2010 for a detailed discussion). In the residential heating market, direct and often single fiscal incentives in the form of e.g. low interest grants, loans, or tax rebates have stimulated installments of wood stoves/boilers (e.g. pellets, wood chips). These incentives were especially effective in markets with mature boiler/stove industries and where the respective boiler fuel was price competitive (on a heating value basis) with other heating fuels. Price prognoses in the residential market although are yet practically non-existent or relatively short-term (Sikkema et al. 2011). Hence, short-term feedstock prices and stove/boiler choices connected to fiscal incentives defined local demand patterns. Trade for the residential market is therefore directly linked to feedstock price advantages. Individual investment support in the power and heating market was only relevant for medium-sized, community-owned projects. Rather, the sector has been critically influenced by regulations (quotas, taxes, feed-in schemes) providing long-term framework conditions. The EU is once again a nice illustration were all aforementioned regulation types have proven to be effective (Held et al. 2010). Quota systems, such as those for renewable electricity production in Belgium, Italy, and the UK, primarily led to large-scale co-/mono-firing in existing power/CHP plants (allowing economies of scale), and thus quota achievements at minimal costs. The taxation of fossil fuels (e.g. coal in Denmark, light heating oil in Italy), partly also connected to quota schemes (e.g. in Sweden) has shown similar effects. Trade patterns were exclusively related to the feedstock flexibility of the combustion technology in order to broaden the sourcing portfolio and level out price fluctuations (Lamers et al. 2012a). This observation has also been made for imports under feed-in schemes, e.g. to the Netherlands, unless they included eligibility criteria. Pulverized coal combustion plants in Belgium, the Netherlands, and UK have limited the utilized streams to wood pellets whereas dedicated grate and/or fluidized bed facilities in Sweden, Italy, or Denmark have sourced a much wider feedstock range including palm kernel shells.

3.6 Conclusions and Outlook

In general, the key constraint for international woody biomass trade for energy is economic viability. Margins are primarily influenced by production and transport costs, but also prices in and exchange rates to target markets. As production costs depend heavily on feedstock prices, it is not surprising that key producing and exporting regions have a long tradition in export oriented forestry, wood processing, and/or pulp and paper industries, and benefit from the availability of low/no cost feedstock and/or residues, infrastructure, and experience.

EU markets are expected to remain the largest driver for international, policy-influenced, woody biomass for energy trade in the near future. In this, the EU will remain a net importer and North America a net exporter of wood pellets, despite the significant domestic consumption of wood pellets in the USA. East Asia is predicted to be-come the second-largest consumer of wood pellets after the EU. The development in Latin America regarding wood pellets is still unclear. Local demand is expected to remain marginal but the region is regarded to have significant production potential (van Vuuren et al. 2009; Beringer et al. 2011; Schueler et al. 2013). Previous investment strategies by Brazilian Suzano Energia Renovavel (subsidiary of paper giant Suzano) alone would see a rise in wood pellet production capacity of 3 Mtonnes by the end of 2015. The pellets are aimed for consumption in Europe. It is yet to be seen however whether such investments take place as wood for pellet production competes with the production of fibre for pulp and paper. Also, Brazilian production would need to qualify under potential future requirements for the sustainability of woody biomass in the importing countries.

Global wood chip trade for energy to the EU is unlikely to increase significantly in the short-term. North American imports underlie phytosanitary measures (due to pine beetle and nematode infection) which increase end prices and limits their use to the higher priced markets such as pulp and paper production. South America, Brazil in particular, and Asia are increasing their pulp and paper production capacities. Previously traded woody biomass, such as chips, will continuously be used within these regions. Africa has slowly increased wood chip production for energy in recent years. Developments are however primarily driven by European companies, and end-use markets will remain off-shore. It is also expected that wood chips are eventually converted to pellets prior to transport.

We know that policy related international trade has risen sharply over the past years. Sophisticated analysis of exact trade volumes however is inherently dependent on robust data and trade codes are still not used consistently on an international scale. Comparisons between officially reported volumes, market data and anecdotal evidence suggest that EUROSTAT statistics provide a good starting basis (especially regarding international wood pellet trade).

References

Alakangas, E., & Keränen, J. (2011). *Report on policy instruments affecting on the forest industry sector and wood availability, survey result report*. Jyväskylä: VTT. Available at: http://www.eubionet.net/GetItem.asp?item=digistorefile;292990;1770¶ms=open;gallery. Accessed 4 July 2011.

Alakangas, E., Heikkinen, A., Lensu, T., & Vesterinen, P. (2007). *Biomass fuel trade in Europe* (ed Eubionet2). Jyväskylä: VTT. Available at: http://eubionet2.ohoi.net/ACFiles/Download.asp?recID=4705. Accessed 5 Apr 2011.

Beringer, T. I. M., Lucht, W., & Schaphoff, S. (2011). Bioenergy production potential of global biomass plantations under environmental and agricultural constraints. *GCB Bioenergy, 3*, 299–312.

Bradley, D., Kranzl, L., Diesenreiter, F., Nelson, R., & Hess, J. (2009). *Bio-trade & bioenergy success stories* (IEA bioenergy task 40). Available at: http://www.bioenergytrade.org/downloads/bioenergysuccessstoriesmar92009.pdf. Accessed 12 Jan 2011.

Cocchi, M., et al. (2011). *Global wood pellet industry and market study* (IEA bioenergy task 40). Available at: http://www.bioenergytrade.org/downloads/t40-global-wood-pellet-market-study_final.pdf. Accessed Jan 2012.

EC. (2010). *Good practice guidance on the sustainable mobilisation of wood in Europe*. Available at: http://ec.europa.eu/agriculture/fore/publi/forest_brochure_en.pdf. Accessed Jan 2013.

Economist. (2012, November 17). It mustn't be business as usual. *Economist*, New York/London/San Francisco.

Edenhofer, O., et al. (Eds.). (2011). *IPCC special report on renewable energy sources and climate change mitigation*. Cambridge/New York: Cambridge University Press.

EUROSTAT. (2013). *Data explorer – EU27 trade since 1995 by CN8*. Brussels, Belgium, Eurostat. Retrieved Jan 2013. Available at: http://epp.eurostat.ec.europa.eu/portal/page/portal/statistics/search_database.

Faber, J., Bergsma, G., & Vroonhof, J. (2006). *Bio-energy in Europe 2005 – Policy trends and issues*. Delft: CE Delft.

FAO. (2012). *2011 Global forest products facts and figures*. Available at: http://www.fao.org/fileadmin/user_upload/newsroom/docs/2011%20GFP%20Facts%20and%20Figures.pdf. Accessed Feb 2013.

FAOSTAT. (2013). *ForesSTAT*. Rome, Italy, Food and Agriculture Organization of the United Nations. Retrieved Jan 2013. Available at: http://faostat.fao.org/site/626/default.aspx#ancor.

FOREST. (2011). *A guide to biomass heating standards*. Rome, Italy: Comitato Termotecnico Italiano Energia e Ambiente. Available at: http://www.forestprogramme.com/files/2011/05/FOREST-Standard-Guide_V04_UK.pdf. Accessed Jan 2013.

Goh, C. S., et al. (2013). Wood pellet market and trade: A global perspective. *Biofuels, Bioproducts and Biorefining, 7*, 24–42.

Goto, S., Oguma, M., Iwasaki, Y., & Hayashi, Y. (2012). *Japan country report 2011* (IEA bioenergy task 40). Available at: http://bioenergytrade.org/downloads/iea-task-40-country-report-2011-japan.pdf. Accessed May 2012.

Guizot, M. (2010). The prospects of wood chips and pellets production and trade. In *Proceedings of the 2nd biomass for heat and power conference,* Brussels.

Heinimö, J. (2008). Methodological aspects on international biofuels trade: International streams and trade of solid and liquid biofuels in Finland. *Biomass and Bioenergy, 32*, 702–716.

Heinimö, J., & Junginger, M. (2009). Production and trading of biomass for energy – An overview of the global status. *Biomass and Bioenergy, 33*, 1310–1321.

Held, A., Ragwitz, M., Merkel, E., Rathmann, M., & Klessmann, C. (2010). *Indicators assessing the performance of renewable energy support policies in 27 Member States*. Available at: http://www.reshaping-res-policy.eu/downloads/RE-Shaping%20D5D6_Report_final.pdf. Accessed Feb 2011.

Hiegl, W., & Janssen, R. (2009). Pellet market overview report – Europe. In *Pellets@las*. Available at: www.pelletsatlas.info. Accessed Feb 2011.

Johnson, F., Tella, P., Israilava, A., Takama, T., Diaz-Chavez, R., & Rosillo-Calle, F. (2010). *What woodfuels can do to mitigate climate change* (FAO Forestry Paper). Available at: http://www.fao.org/docrep/013/i1756e/i1756e00.pdf. Accessed Mar 2011.

Junginger, M., van Dam, J., Alakangas, E., Virkkunen, M., Vesterinen, P., & Veijonen, K. (2010). Solutions to overcome barriers in bioenergy markets in Europe. In *EUBIONET3*. Available at: http://www.eubionet.net/GetItem.asp?item=digistorefile;144551;1087¶ms=open;gallery. Accessed Jan 2011.

Kaltschmitt, M., & Hartmann, H. (2001). *Energie aus Biomasse: Grundlagen, Techniken und Verfahren*. Berlin: Springer.

Lamers, P., Junginger, M., Hamelinck, C., & Faaij, A. (2012a). Developments in international solid biofuel trade – An analysis of volumes, policies, and market factors. *Renewable and Sustainable Energy Reviews, 16*, 3176–3199.

Lamers, P., Marchal, D., Schouwenberg, P. P., Cocchi, M., & Junginger, M. (2012b). *Global wood chip trade for energy* (IEA bioenergy task on 40 Sustainable International Bioenergy Trade). Available at: http://www.bioenergytrade.org/downloads/t40-global-wood-chips-study_final.pdf. Accessed June 2012.

Lamers, P., Thiffault, E., Paré, D., & Junginger, H. M. (2013). Feedstock specific environmental risk levels related to biomass extraction for energy from boreal and temperate forests. *Biomass and Bioenergy, 55*(8), 212–226.

Mitchell, C., et al. (2011). Policy, financing and implementation. In O. Edenhofer, R. Pichs-Madruga, Y. Sokona, K. Seyboth, P. Matschoss, S. Kadner, T. Zwickel, P. Eickemeier, G. Hansen, S. Schlömer, & C. V. Stechow (Eds.), *IPCC special report on renewable energy sources and climate change mitigation*. Cambridge/New York: Cambridge University Press.

Olsson, O., Vinterbäck, J., Dahlberg, A., & Porsö, C. (2010). *Price mechanisms for wood fuels*. Available at: http://www.eubionet.net/GetItem.asp?item=digistorefile;178383;1087¶ms=open;gallery. Accessed June 2011.

Ragwitz, M. et al. (2007). OPTRES – *Assessment and optimisation of renewable energy support measures in the European electricity market*. Intelligent Energy for Europe – Programme (Contract No. EIE/04/073/S07.38567). Available at: http://www.optres.fhg.de/OPTRES_FINAL_REPORT.pdf. Accessed Feb 2011.

REN21. (2012). *Renewables global status report*. Renewable Policy Network for the 21st century. Paris: REN21. Available at: http://www.ren21.net/default.aspx?tabid=5434. Accessed Jan 2013.

Schueler, V., Weddige, U., Beringer, T., Gamba, L., & Lamers, P. (2013). Global biomass potentials under sustainability restrictions defined by the European Renewable Energy Directive 2009/28/EC. *GCB Bioenergy*, online early view, doi:10.1111/gcbb.12036.

Sikkema, R., Steiner, M., Junginger, M., Hiegl, W., Hansen, M., & Faaij, A. (2011). The European wood pellet markets: Current status and prospects for 2020. *Biofuels, Bioproducts and Biorefining, 5*, 250–278.

Smeets, E., Faaij, A., & Lewandowski, I. M. (2007). A bottom-up assessment and review of global bio-energy potentials to 2050. *Progress in Energy and Combustion Science, 33*, 56–107.

UN. (2013). *Commodity Trade Statistics Database (COMTRADE)*. New York, Geneva, United Nations. Retrieved Jan 2013. Available at: http://comtrade.un.org/db/default.aspx.

UNECE/FAO. (2010). *Forest product conversion factors for the UNECE region*(Geneva Timber and Forest Discussion Paper 49). Geneva: United Nations Economic Commission for Europe, Food and Agriculture Organization of the United Nations. Available at: http://www.unece.org/fileadmin/DAM/timber/publications/DP-49.pdf. Accessed Feb 2013.

Uslu, A., Faaij, A., & Bergman, P. (2008). Pre-treatment technologies, and their effect on international bioenergy supply chain logistics. Techno-economic evaluation of torrefaction, fast pyrolysis and pelletisation. *Energy, 33*, 1206–1223.

van Vuuren, D., van Vliet, J., & Stehfest, E. (2009). Future bio-energy potential under various natural constraints. *Energy Policy, 37*, 4220–4230.

Vikinge, B. (2011). Trading of wood chips in the Baltic Sea region. In *EUBIONETIII workshop: Biomass trade – focus on solid biofuels*, Espoo.

Willumsen, K. (2010). Import of wood-chips from non-European countries – need for a common PRA? In *Nordic-Baltic plant health meeting*, Finland.

Chapter 4
Development of Bioenergy Trade in Four Different Settings – The Role of Potential and Policies

Daniela Thrän, Christiane Hennig, Evelyne Thiffault, Jussi Heinimö, and Onofre Andrade

Abstract The provision, use and trade of bioenergy differ significantly between countries. This chapter provides an overview of bioenergy trade worldwide and presents case studies of four national biomass markets – Brazil, Canada, Finland and Germany – showing diverging degrees of biomass use for energy provision and biomass potentials. Since energy policy is considered to be a main driver for the use of biomass for energy generation, an overview of bioenergy policy making in different countries and the resulting impact on trade is given.

Today, dedicated solid biomass and liquid biofuels are the most relevant traded commodities. The expected stronger demand for biomass resources in particular in IEA Bioenergy Task 40 countries by 2020 may induce further trade activities. The majority of the OECD countries has implemented different support schemes pri-

D. Thrän (✉)
Bioenergy Systems, Deutsches Biomasseforschungszentrum gGmbH – DBFZ, Helmholtz-Zentrum für Umweltforschung GmbH – UFZ, Leipzig, Germany
e-mail: daniela.thraen@ufz.de

C. Hennig
Bioenergy Systems, Deutsches Biomasseforschungszentrum gGmbH – DBFZ, Leipzig, Germany
e-mail: christiane.hennig@dbfz.de

E. Thiffault
Canadian Forest Service – Natural Resources Canada, Government of Canada, Quebec, Canada
e-mail: evelyne.thiffault@rncan-nrcan.gc.ca

J. Heinimö
Faculty of Technology, Lappeenranta University of Technology, Varkaus, Finland
e-mail: jussi.heinimo@lut.fi

O. Andrade
Department of Biofuels, Argos, Rotterdam, The Netherlands
e-mail: onofre.andrade@argosenergies.com

M. Junginger et al. (eds.), *International Bioenergy Trade: History, status & outlook on securing sustainable bioenergy supply, demand and markets*, Lecture Notes in Energy 17, DOI 10.1007/978-94-007-6982-3_4,

marily for liquid biofuels and power generation from biomass and has set ambitious political targets for bioenergy for the coming decade. Data availability regarding the resource situation is one precondition for a clear target definition for the development of the bioenergy sector. However, this information is often difficult to obtain.

The role of trade in the different countries depends especially on the political support and the specific resource situation, taking into account the overall biomass potential and the expected domestic demand. Hence, for some countries resource limitations can be expected to become relevant when implementing the set national targets for bioenergy. Consequently, this might lead to a comparable higher future import demand for biomass in many OECD countries.

4.1 Introduction – Role of Bioenergy in Different Countries

The provision, use and trade of bioenergy differ significantly between countries. This is mainly due to the type and amount of available energy resources and the energy policy objective of a country.

By the means of a country's biomass potential and the relation between biomass potential and use of biomass for energy provision, differences in the role of bioenergy and the prospects for bioenergy trade between countries can be described. An overview of those two indicators for IEA Bioenergy Task 40 countries plus China, Russia and India is presented in the figure below. For the evaluation the years 2004 and 2020 have been considered.

The assumed technical biomass potential in the different countries refers to a "bioenergy scenario" implying a strong and substantial energy-related biomass use (DBFZ 2011). For a sound comparison of the studied countries the biomass potential is put in relation to the number of inhabitants (potential per capita). In terms of biomass use both data from national statistics offices and from the International Energy Agency have been applied.

As Fig. 4.1 shows, all IEA Bioenergy Task 40 countries aim at an increasing use of biomass for energy provision until 2020 in comparison to the reference year 2004 (reflected by the arrows). On the contrary, countries like India, China and Russia are not focusing on a larger share of bioenergy within the scope of their energy policy. The per capita biomass potential in the considered countries differs even in a wider range (reflected by the bars). It can be concluded that trade in immature markets does also strongly depend on policies. A stronger demand for biomass resources in the future may induce further trade activities since most of the countries are going to deplete or even exceed their domestic biomass potential for energy by 2020 (indicated by the dotted line).

The following sections describe the bioenergy trade in different countries considering both an overview of the most relevant countries and insights in four national biomass markets to illustrate characteristics, similarities and differences. Based on that, country case studies have been chosen representing a broad range of biomass potentials and biomass uses. Since energy policy is considered to be a main driver for the use of biomass for energy generation, an overview of bioenergy policy making in the different countries and the resulting impact on trade is given beforehand.

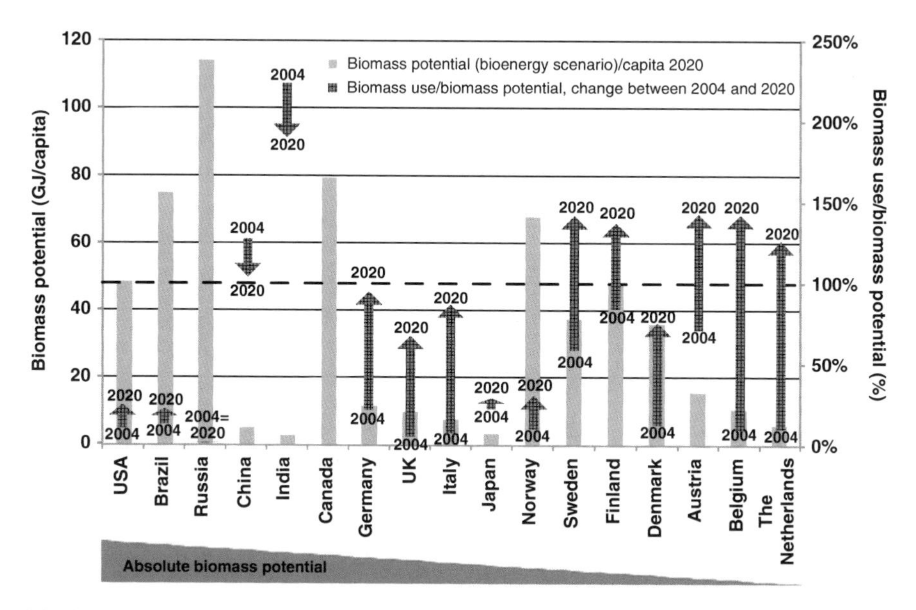

Fig. 4.1 Specific biomass potential for energy per capita and biomass use in selected countries (No projections on the biomass use in Canada by 2020 have been available) for 2004 and 2020. *The countries are characterized by a decreasing absolute biomass potential from the *left* to the *right* (Source: DBFZ 2011; IEA 2006a, b, 2011; Thraen et al. 2012; Walter and Dolzan 2012; Rosillo-Calle and Galligani 2011; Tromborg 2011; Hektor 2012; Nikolaisen 2012; Kalt et al. 2011; Guisson and Marchal 2011; Goh et al. 2012; Cocchi 2012; Heinimö and Alakangas 2011a)

4.2 Imports and Exports of Different Countries

As presented in Chaps. 2 and 3 of this book the most significant biomass resources that are traded via long distances are bioethanol, biodiesel, wood pellets and wood chips. The trading flows and corresponding volumes for wood pellets and biodiesel for an energy related use can be assigned directly. However, in the case of wood chips and bioethanol the trading volumes intended for energy generation are aggregated with the traded volume for industrial and other applications. Hence only the total quantity can be reported. In Fig. 4.2 the imported and exported quantities within the IEA Bioenergy Task 40 countries in 2010 are presented. Taking into account wood pellets and biodiesel as energy commodities, it can be pointed out that the IEA Bioenergy Task 40 countries cover a major part of the traded biomass. About 52 % of the global production volume of wood pellets (14.3 million tons in 2010) has been demanded via trade by this country group. With regard to biodiesel, 25 % of the global production volume (16.6 million tons in 2010) can be assigned. Thus these countries are the main drivers for trade of biomass for energy use (F.O. Licht's 2011; Cocchi et al. 2011).

With regard to the country case studies two exporters and two importers have been considered.

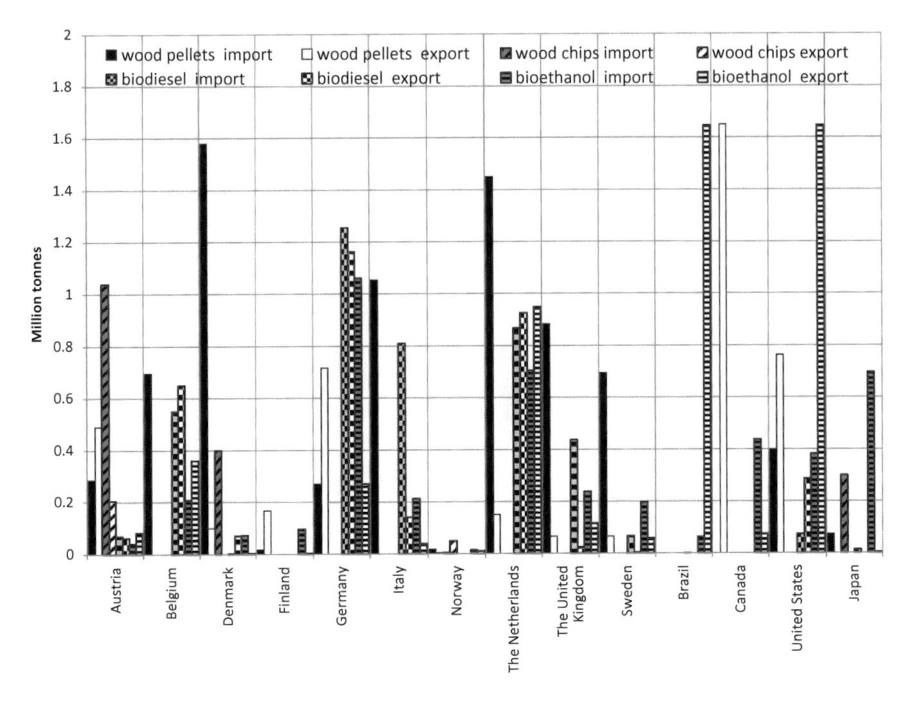

Fig. 4.2 Biomass import and export in IEA Bioenergy Task 40 countries in 2010 (for wood chips and bioethanol aggregated data) (F.O. Licht's 2011; Cocchi et al. 2011)

4.3 Bioenergy Policy

Since the use of renewable energy sources has been recognized as a vital component in combating climate change, in many countries the direction and design of energy policy frameworks have been adapted in favour of promoting RES, leading to ambitious targets for realizing a larger share in the overall energy mix. Also for the provision, use and trade of bioenergy the specific political framework conditions are very relevant. In the following section the bioenergy policy framework in selected IEA Bioenergy Task 40 countries has been analysed and assessed according to its impact on biomass use for energy generation and related trade of biomass.

Table 4.1 gives an overview of the policy and regulatory framework for promoting energy-related use of biomass in the various IEA Bioenergy Task 40 countries around the world. The policy instruments[1] are grouped according to financial and regulatory measures. These measures differ in their degree of enforcement as

[1] Classification of public policy instruments: financial (investment subsidy, operational subsidy, tax incentive), regulatory (building regulation, quota obligation), standards (fuel quality standard, sustainability standard) (compared with Bemelmans-Videc et al. 2010).

Table 4.1 Bioenergy policy characteristics and framework for selected IEA Bioenergy Task 40 countries

		Policy framework	
Country	Bioenergy policy characteristics	Financial instruments	Regulations
Denmark	Focal point: use of biomass in large-scale installations Support of biomass use for cogeneration and district heating Co-firing of biomass is supported for generating renewable electricity	Feed-in tariff system: subsidy for electricity produced by combusting biomass (also co-firing) – **operational subsidy** **Tax exemptions** for biomass	The utilities are forced by government decree to use biomass in large amounts
Germany	Focal point: promotion of energy-related biomass use in small- to medium-scale heat and/or power installations No co-firing of biomass is supported for generating renewable electricity Shift from fiscal instruments towards more regulations in the recent years Bioenergy policy on regional and national level	Market incentive programme – **investment subsidy** Renewable Energy Sources Act (Feed-in Tariff) for electricity from biomass – **operational subsidy** **Tax exemptions and deductions** for certain biofuels	Renewable Energies Heat Act (EEWärmeG) – **building regulation** Biofuels Quota Act (BioKraftQuG) – **quota obligation**
Italy	Focal point: promotion of energy-related biomass use in small- to medium-scale heat and/or power installations No co-firing is supported for generating renewable electricity In particular support regime for small scale heat production and energy efficiency Application of both feed-in tariff and green certificates	Feed-in tariff for plants up to 1 MW_{el} which produce and sell electricity from biomass – **operational subsidy** Green certificates for plants that are bigger than 1 MW_{el} – **operational subsidy** **Investment subsidies** for the installation of biomass heating systems	Mandatory integration of biomass heating systems in new buildings or old buildings subject to major renovations – **building regulation** Financial Law 2007: mandatory quota for biofuels used in conventional transport fuels – **quota obligation**

(continued)

Table 4.1 (continued)

Country	Bioenergy policy characteristics	Policy framework: Financial instruments	Regulations
Belgium	Policy is divided in the regions Wallonia, Flanders and Brussels resulting in differing designs of policy Support of co-firing for electricity generation Focal point: Promotion of electricity from biomass In particular use of fiscal instruments for the promotion of bioenergy	In all regions green certificates for RES electricity production are implemented – **operational subsidy** In all regions investment subsidy for the use of biomass for heating – **investment subsidy** Tax deduction for the provision of heat and electricity from biomass and tax exemptions for certain biofuels – **tax incentives**	**Quota obligation** for biofuels used in conventional transport fuels
The Netherlands	Strong support of co-firing of biomass in coal-fired power plants Various deals between government and business about concrete projects in the areas energy saving, renewable energy, sustainable mobility, etc. were signed	The Stimuleringsregeling Duurzame Energieproductie SDE+: subsidy for electricity, heat and gas sector – **investment subsidy**	The Nederlands Beileid Biobrandstoffen: mandatory share of biofuels in the transport sector – **quota obligation**
The United Kingdom	Strong support of co-firing of biomass in coal-fired power plants Electricity market reform Shift from green certificates to feed-in tariffs	Renewable Heat Incentive (RHI) for non-domestic installations – **operational subsidy** Renewable Heat Premium Payment (RHPP) for domestic installations – **investment subsidy** Feed-in Tariff – **operational subsidy** Renewable Obligation Certificates (ROCs) – **operational subsidy**	Renewable Transport Fuel Obligation (RTFO) – **quota obligation**

(continued)

Table 4.1 (continued)

Country	Bioenergy policy characteristics	Policy framework	
		Financial instruments	Regulations
Canada	Dynamic bioenergy policy development at the provincial and/or regional level Renewable Fuels Strategy: sets requirements for fuel producers and importers EcoEnergy Innovation Initiative: boost renewable energy supplies and develop cleaner energy technologies, including bioenergy Support and promotion of forest bioenergy as part of the renewal of the forest industry – cogeneration or community heating projects	In some provinces or at the federal level: **Carbon tax** applied on the purchase or use of liquid fuels and/or coal Credit program for establishment of projects for heat, electricity or cogeneration derived from biomass, and for biofuel production – **operational subsidy** Tax credit for the production of biofuels (per liter of biofuel produced) – **tax incentive**	Requirements, both at the federal level and in some provinces, on the inclusion of a share of renewable fuels in liquid fuels – **quota obligation**
United States	The Energy Independence & Security Act of 2007 (EISA) sets goals for biofuel production through 2022 Emission compliance strategies at the state level have also started to actively enforce Renewable Portfolio Standards (RPS) Other organizations have set targets that while not mandatory, have helped drive federal policy	Renewable Energy Credit (REC) program for production of power from RES (REC's under state sponsored RPS) – **operational subsidy** Production Tax Credit provides a 2.2 cent per kilowatt-hr benefit for the first 10 years of a renewable energy facility's operation – **tax incentive**	RFS-2 sets **quota obligations** for biofuel production from biomass through 2022 State led RPS set **quota obligations** for electricity production from biomass

(continued)

Table 4.1 (continued)

Country	Bioenergy policy characteristics	Policy framework	
		Financial instruments	Regulations
Norway	Doubling of domestic bioenergy production by 2020 decided in 2008 Bioenergy policy centred on heat production due to extensive hydro power production Liquid biofuel policies on hold	**Investment subsidy** for heat production from biomass Green certificate for new renewable electricity production – **operational subsidy** Production subsidies for biomass production Carbon tax on fossil fuels	Biofuels directive not yet decided
Brazil	Use of biofuels focus on securing supply for internal market and making it an international commodity (ethanol) and having a policy directed to diversification of feedstock (today around 80 % soy oil) (biodiesel) Better use of sugarcane bagasse/leafs for energy production and advanced biofuels; modernization of existing production capacity; better grid connectivity for ethanol mills An overarching program Fundo Clima promotes energy from biomass and other renewable (with the exception of sugarcane bagasse)	In 2012, the "Gasoline Tax" was removed, making ethanol less competitive with gasoline **Tax benefits** for biofuels based on "Social Label"; Fundo Clima: low rates for investment loans – **investment subsidy**	Mandate for Anhydrous (added to Gasoline) of 20 % and for biodiesel 5 % (B5) – **quota obligation** Regulations for commercialization of electricity from Bagasse: 1. sales of electricity to distribution companies via auction, 2. direct sales of electricity to consumers without need for auction Regulation concerning micro-generation (until 100 kW) and mini-generation (from 100 kW to 1 MW) for renewable sources

(continued)

Table 4.1 (continued)

		Policy framework	
Country	Bioenergy policy characteristics	Financial instruments	Regulations
Finland	Focal point: promotion of forest chip use in heat and CHP production Biomass is the most important source of renewable energy in the country Majority of bioenergy production and use takes place in forest industry Bioenergy policy and policy instruments are uniform in all regions of the country	Tax relief for all fuels used for electricity generation – **tax incentive** Feed-in tariff for wood fuel based small-scale CHP – **operational subsidy** **Investment subsidy** for biomass **Energy taxation** for fossil fuels used for heat generation	Obligation to distribute biofuels to the transport market
Austria	Focal point: promotion of energy-related biomass use in small- to medium-scale heat and/or power installations Bioenergy strategy on the increase of the share of cogeneration and conversion efficiency as well as use of biogas for electricity, heat and transportation Bioenergy policy on regional and national level	Green Electricity Act (Feed-in Tariff) – **operational subsidy** Support for biomass heating systems and micro-grids in industrial and commercial buildings – **investment subsidy** Agreement acc. Art.15a B-VG (support for biomass heating systems in residential buildings) – **investment subsidy** **Tax reduction and exemption** for certain biofuels	Biofuels Directive – **quota obligation**
Sweden	Strong use of fiscal instruments in particular energy taxation Focal point: promotion of energy-related biomass use in small- to medium-scale heat and/ or power installations No co-firing of biomass is supported for generating renewable electricity	**Tax exemption** for biofuels **Carbon dioxide tax** for fossil fuels Green certificate system for electricity from biomass – **operational subsidy** **Investment subsidies**	

well as market intervention by governmental institutions. Regulations define and ask for a certain behaviour from the different market actors whereas financial measures set incentives to either pursue or to refrain from taking a certain action (Bemelmans-Videc et al. 2010).

The evaluation of the energy policy framework in the various countries shows that in all of the countries measures for the promotion of bioenergy are in place. Main attention of the countries' energy strategies has been on biomass use in the electricity and the transport sector during the last decade. However, in recent years the support of bioenergy in the heating sector came to the fore.

Within the power sector feed-in tariff systems and green certificate systems are the main policy instruments in order to achieve a larger share of renewables among the countries. To achieve an increased use of biofuels, mandatory quota systems – a regulatory instrument – are applied. In contrast biomass for heating is mainly promoted through financial subsidies reducing the initial investment of a heating installation. Hence, many countries rely on providing incentives in order to attain a certain behaviour. In a few countries a shift towards building regulations could be observed implying that the use of renewables for heating in buildings is mandatory.

As a result energy policies with a focus on promoting bioenergy led and will lead to an increased demand for biomass resources. Depending on the national or regional potential this has already induced wider trade of biomass and a further expansion of the trading network can be expected in the years to come. Next to the trade intensity also the type of feedstock traded is strongly influenced by the national energy policy.

For example in the European Union the Renewable Energy Directive (2009/28/EC)[2] sets mandatory targets for the use of RES for energy provision until 2020 and the corresponding realization by means of national action plans. The overall share of renewable sources in the EU's total energy mix shall amount to 20 % and within the transportation sector to at least 10 % (Directive 2009/28/EC 2009). In the majority of the national Renewable Energy Action Plans (nREAP) of EU countries the achievement of the 2020 objectives is linked to a larger share of bioenergy. Thereby most of the countries consider an increase in biomass imports.

Besides, an amendment of the Renewable Energy Directive is currently under consideration. This amendment includes the definition of the type of feedstock used for achieving the 10 % target in the transport sector. Only 5 % of this target can be achieved based on biofuels of the first generation and the remainder comes e.g. from wastes and residues.[3] All these aspects determine the type and quality of biomass traded and therefore will influence the development of future trading flows.

[2] Directive 2009/28/EC of 23 April 2009 on the promotion of the use of energy from renewable sources and amending and subsequently repealing Directives 2001/77/EC and 2003/30/EC.

[3] COM(2012)595, Proposal for a Directive amending Directive 98/70/EC relating to the quality of petrol and diesel fuels and amending Council Directive 93/12/EC and amending Directive 2009/28/EC on the promotion of the use of energy from renewable sources, 595 p.

4.4 Country Case Studies

4.4.1 Germany

4.4.1.1 General Introduction

Germany is a country with a high population density and a strong energy demand from different industries. In 2011 the government decided to accelerate the transition of the energy sector towards renewable sources and is currently facing different challenges especially in the electricity sector (BMU 2013).

In 2010 the total final energy consumption in Germany amounted to 9,060 PJ (BMU 2011). Germany has little indigenous fossil energy resources. The only significant source has been coal over the last century. However its potential is mostly developed and the energy production is not cost competitive anymore. Today, the most important sources used for energy provision are mineral oil (33.3 %) and natural gas (21.9 %) which are mainly imported (AGEB e.V 2011). Between 1990 and 2010, the share of renewable energy within the German energy system has increased more than fivefold – from 2 to 11 % of the final energy consumption (BMU 2011). Among the different energy sectors renewable energies have the largest share within the electricity sector amounting to 16.8 %. The contribution of renewable energy sources (RES) to heating and transportation has been 9.8 and 5.8 %, respectively, in 2010.

With a share of 7.7 % of the total final energy consumption biomass presents 70 % of the renewable energy sources in Germany (Fig. 4.3). Concerning the different energy sectors biomass is the dominating renewable energy source in the transport and heating sector and contributes a vital share to the electricity sector (30.5 % of RES) (AGEB e.V 2011).

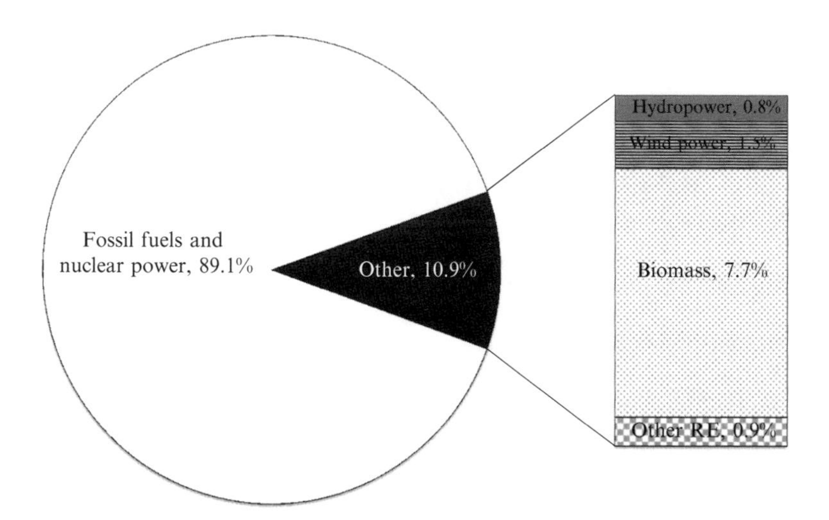

Fig. 4.3 Total final energy consumption in Germany in 2010 (BMU 2011)

The German national renewable energy action plan (nREAP) of 2010 includes the following targets for the different energy sectors by 2020: 30 % of RES for electricity generation, 14 % of RES for heating and cooling and 10 % of RES for transportation. Thereby biomass shall contribute with 22.8 % to the electricity sector, 78.7 % to the heating and cooling sector and with almost 100 % to the transportation sector (reference scenario). By 2020 the share of biomass is expected to amount to almost 10 % of the total final energy consumption (8,859 PJ) in Germany (Federal Republic of Germany 2010). For achieving these objectives the necessity of importing biomass has been indicated due to a lack of resources and too short timeframe for unlocking untapped biomass potential.

4.4.1.2 Biomass for Energy – Production and Consumption

The main resources for the provision of bioenergy are waste streams (waste wood, biodegradable waste, yellow grease etc.), residues from agriculture and the food-processing industry (manure, distiller's wash etc.), residues from forestry and the wood-processing industry as well as energy crops. In 2010 energy crops were grown on 1.8 million hectare, mainly rapeseed oil for the biodiesel production and provision of vegetable oil (51 %) (FNR 2013). At the end of 2010 approximately 5,900 biogas plants and 311 biomass power plants fed electricity into the national grid (DBFZ 2012a) and 30 biodiesel plants and 8 bioethanol plants produced liquid biofuels for transportation (DBFZ 2012b). Additionally there is a strong interest in upgrading biogas to biomethane and feeding it into the national gas grid.

Within the heating sector biomass is mainly used in small- to medium-scale installations. Hence, there is a big market of biomass boiler and stoves for individual heat supply – mainly from local re-sources.

The most important woody biomass resources for energy generation are firewood, wood pellets and wood chips in Germany. In the following the focus of the market description is on wood pellets since this is the woody biomass commodity that is significant for trading over long-distances and is expected to have high market relevance in the future.

Wood pellets are used in small- to medium-scale installations for heating purposes in Germany. From 2004 until 2010 both consumption and production have increased constantly (Fig. 4.5). This development has been induced by the national policy support. Next to the establishment of a national wood pellet market also outside of Germany markets have evolved. Due to the fact that Germany has a substantial wood-processing industry with corresponding residue volumes and an increasing demand for wood pellets also outside Germany, the national production capacities have increased in the past. Today the production volume is exceeding the consumption significantly and Germany is a net exporter of wood pellets. In the coming years a consolidation of the domestic production and capacities is expected since the Europe wide and even worldwide competition has intensified whereas at the same time the national consumption is stagnating.

Among liquid biofuels the use of biodiesel in the transport sector is most prominent in Germany. Since 2004 a substantial rise in the demand could be observed until

2007. From 2008 onwards the consumption has been stagnating. One reason for this market development is that the policies promoting its use have not been pursued consistently enough. In recent years, also the production and consumption of bioethanol as fuel could be established and has increased significantly, even though its market share is still far below that of biodiesel (compare Fig. 4.5).

The production and consumption of biomethane is seen as an important part for the transition of the energy system towards renewable sources and will become more significant in the next years (Fig. 4.4).

4.4.1.3 Biomass for Energy – Prices

The price for wood pellets (ENplus quality) has been fairly steady with price levels of 160 to 180 €/ton (excl. VAT) till 2005. In 2006 the price started to rise with a peak of 245 €/ton (excl. VAT) in December. This has been due to a sudden increase in demand in Europe with a short supply. In the following years the price has stabilized. During 2010 the price for wood pellets ranged between 204 and 219 €/ton (excl. VAT) (DEPV e.V. 2011b; C.A.R.M.E.N. e.V. 2011a). For the next years a stabilization of the market prices can be expected since the market is going to consolidate.

The price for wood chips from forest residues used for heat generation in small- to medium-scale installations has steadily increased over time. Until 2010 a price increase of 100 % could be observed (reference year 2004). Reasons for this development are the higher demand for wood chips as energy commodity and also the general higher price level of fossil fuels which is often considered as reference for the price development of wood chips. In 2010 the average prices for wood chips ranged between 76 and 85 €/ton (excl. VAT) (C.A.R.M.E.N. e.V. 2011b). In future a further price increase is expected due to a higher demand but also an overall increase in the energy price level.

The prices for biodiesel (end consumer prices) have been stable in Germany in the years 2005, 2006 and 2007. In 2008 prices have experienced a tremendous growth which was due to the increased prices for the raw material vegetable oil (palm oil and rapeseed oil). During that year the biodiesel price has increased by 50 % compared to the reference year 2005. In 2009 the price returned to the 2005/2006 level. During 2010 the average price was 960 €/ton (excl. VAT) (F.O. Licht's 2011).

After a price drop in 2006 the price level of bioethanol has again increased in Germany until 2010. From 2006 to the end of 2010 the price level has risen by about 60 %. The average price amounted to 362 €/t (excl. VAT) in 2010 (F.O. Licht's 2011).

The following graph (Fig. 4.5) gives an overview on the price development of bioenergy commodities in Germany between 2004 and 2010.

4.4.1.4 Biomass for Energy – Trade

In Germany the major traded volumes for biomass resources can be recorded for fuel wood, wood pellets, biodiesel and bioethanol. About 30 % of the produced wood pellets in 2010 (1.75 million tons) have been exported mainly for the use in

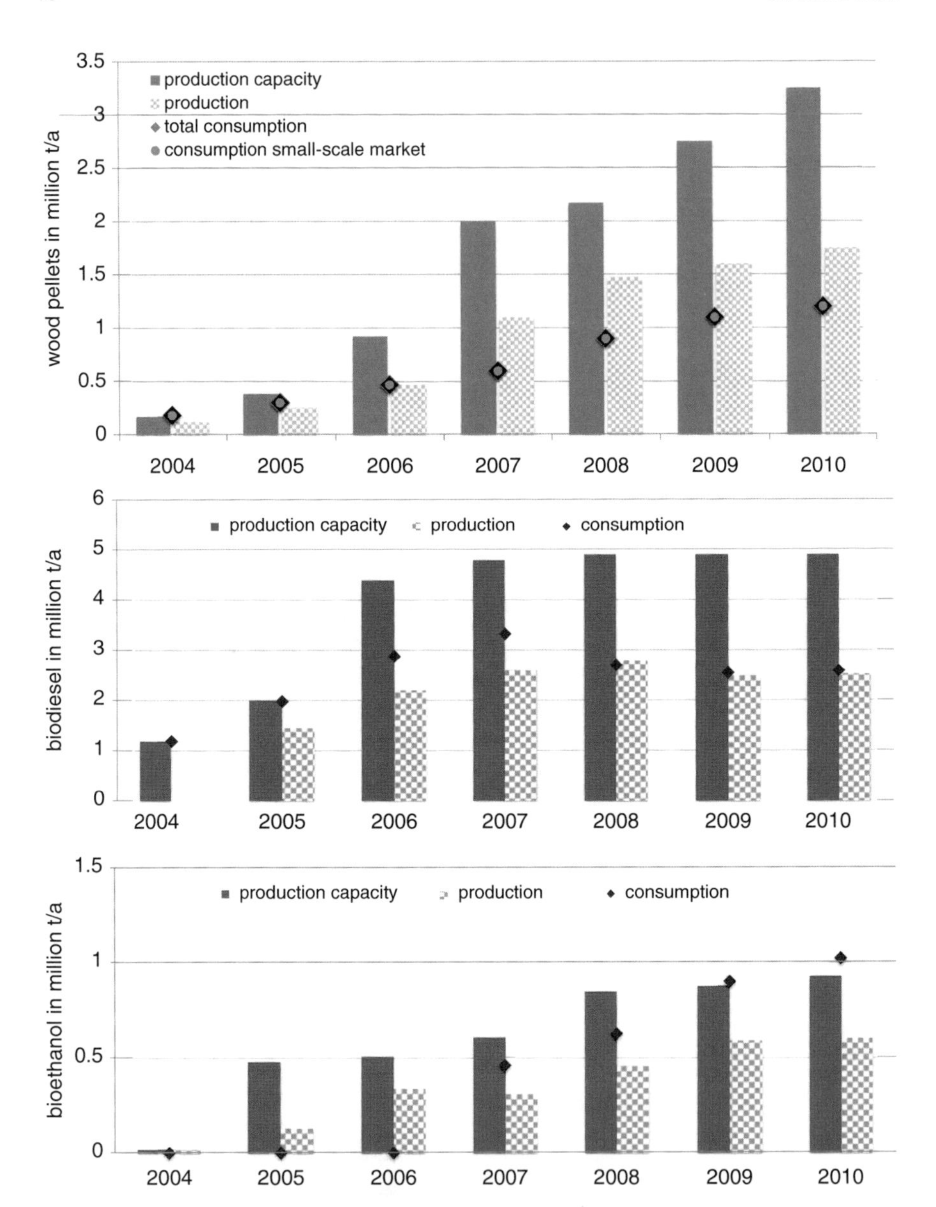

Fig. 4.4 Production capacity, production and consumption of wood pellets, biodiesel and bioethanol in Germany, 2004–2010 (Source: DEPV e.V. 2011a; Solar Promotion GmbH 2010; VDB 2011; BAFA 2011; F.O. Licht's 2011)

co-incineration installations. On the imports side both certified wood pellets for heating purposes and industrial pellets were traded from Austria, Czech Republic, the Baltics and Belarus to Germany. Thereby the industrial pellets are fully re-exported. In the past trade has been particularly over rather short distances and within

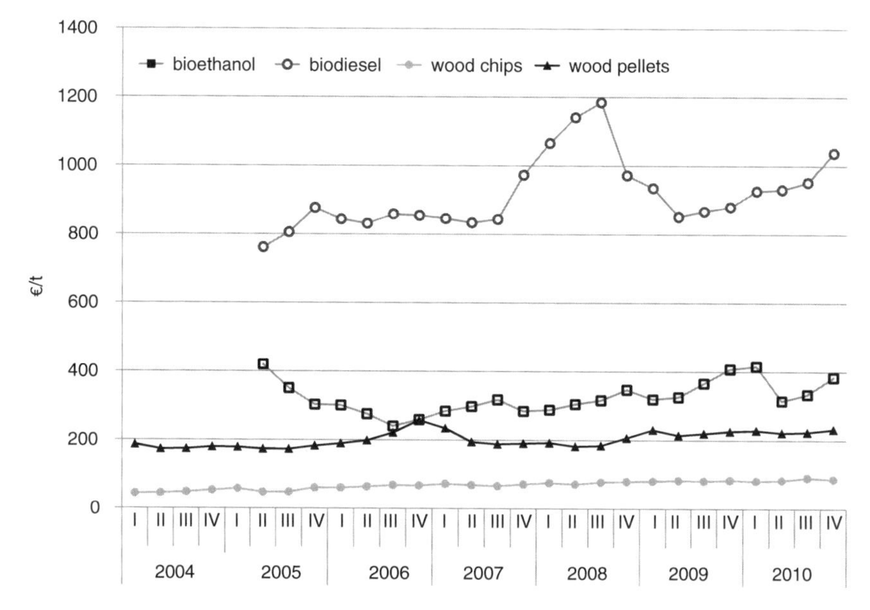

Fig. 4.5 Prices for wood pellets, wood chips, biodiesel and bioethanol in Germany, 2004–2010 (excl. VAT) (DEPV e.V. 2011b; C.A.R.M.E.N. e.V. 2011a, b; F.O. Licht's 2011; BDBE e.V. 2009)

bordering countries (Thrän et al. 2012). Recently trading streams from further Eastern European countries have been induced, which is especially due to improved qualities of the commodities at lower prices and an increasing demand for wood pellets in Europe. At the same time pellet producers face a higher share of unused production capacities. In the near future increasing trading activities on the German market are not to be expected, since the German as well as European market have entered a consolidation phase and suppliers from Eastern Europe can provide wood pellets at lower prices.

Looking at the biodiesel and bioethanol market, trading activities are rather global. The main trading partners for bioethanol were the United States of America, Brazil and Canada (via the port of Rotterdam) in 2010. It has to be noted that the trading volumes intended for energy generation are aggregated with the traded volume for industrial and other applications. Hence in many cases only the total quantity is reported. Germany is a net importer of bioethanol. Next to bioethanol also biodiesel is handled via the port of Rotterdam. Argentina and Indonesia are the main exporting countries from overseas to Germany. The biodiesel production based on rapeseed is well developed domestically. Thus also a significant share of the production volume has been exported to Poland, the Netherlands, Belgium and France in 2010. Overall Germany is a net importer of biodiesel (Thrän et al. 2012).

For the trade of biomethane, requirements and regulations are currently being defined and established; but there are still a lot of impediments for an international trade, i.e. technical standards, demands on the gas quality, balancing rules (Fraunhofer UMSICHT 2012).

4.4.2 Canada

4.4.2.1 General Introduction

Canada is the world's second-largest country by area at over 9 million km^2. It has a population of over 34 million people and is made up of ten provinces and three territories. It has considerable natural resources including oil and gas, coal, hydro, minerals, and forests.

There are approximately 397.3 million hectares of forest and other wooded lands within Canada, which represent 10 % of the global forest cover and 30 % of the world's boreal forest cover (Natural Resources Canada 2011). 229 million hectares of forest in Canada is considered to be under management. As of December 2010, 149.8 million hectares of managed forest were certified as being sustainably managed by one or more of globally recognized certification standards (Natural Resources Canada 2011).

Under Canada's constitution, the federal government and the provinces/territories have specific roles in the care and governance of natural resources and public forests, as well as sharing responsibility for matters such as environmental regulation and science and technology. The provinces and territories collectively own 77 % of Canada's forested area, while the federal government owns 16 % and private land owners control the remaining 7 %. The country has a well-developed forest sector and has historically been one of the world's largest exporters of wood products. This sector makes up at least 50 % of the economic base for approximately 200 Canadian communities (Natural Resources Canada 2011).

Agricultural farmland comprised 6.4 million km^2 in Canada in 2011 (down 4.1 % since the last census in 2006), or around 7 % of the total land base (Statistics Canada 2012). Crops are grown on 53.1 % of farmland. In 2010, the Canadian agriculture and agrifood system accounted for 8.1 % of total GDP and provided one in eight jobs. Although primary agriculture represents a small share of the total economy, it has grown an average of 1.5 % per year since 1995. In 2010, Canada was the world fifth largest exporter of agricultural goods, and the world sixth largest importer (Agriculture and AgriFood Canada 2012)

Canada's energy sector accounted for 6.8 % of GDP in 2010. Canada is one of the few developed economies being a net exporter of energy with the gap between exports and imports growing. In 2010, net energy exports amounted to $CAN 48 billion, with crude oil, petroleum and coal representing 72 % of net export revenue (Canada National Energy Board 2011a).

The share of renewable energy within the Canadian energy production portfolio is about 11 % since 2007; of that share, hydropower counts for 8 %, whereas biomass represents about 3 % (compare Fig. 4.6) (Canada National Energy Board 2011a). At the end of 2010, Canada had 61 bioenergy power plants with a total installed capacity of 1,700 MW. Most of this capacity was built around the use of wood biomass and spent pulping liquor, as well as landfill gas. In 2010, 8.3 GW hours of electricity were generated using wood refuse and spent pulping liquor, representing 1.4 % of the total electricity generated in Canada (Natural Resources

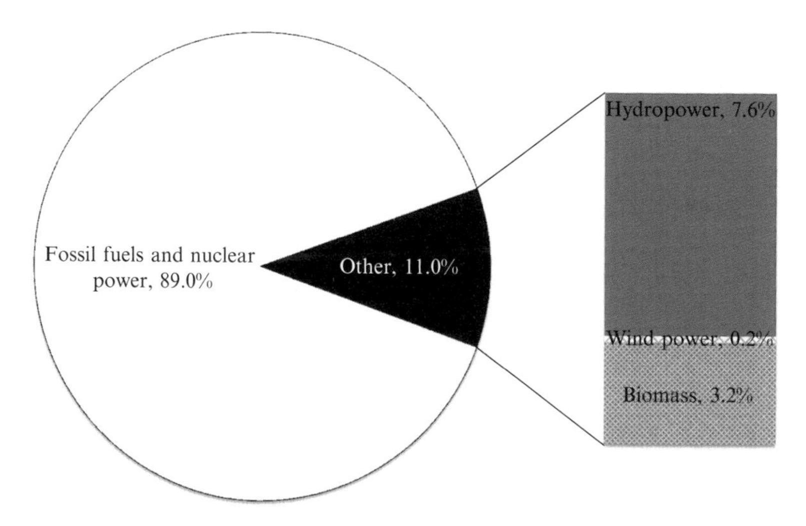

Fig. 4.6 Total final energy consumption in Canada in 2010 (Canada National Energy Board 2011a)

Canada 2012). In 2010, 7 % of total energy consumption for the residential, commercial and industrial sectors came from biomass (for the direct production of heat), geothermal and solar sources. Almost all of the bioenergy used in the Canadian industrial sector is attributable to the pulp and paper industry. In 2009, bioenergy accounted for 58 % of the total energy used by the forest industry, with the pulp and paper sector meeting 62 % of its energy needs from forest biomass (Natural Resources Canada 2011). In the transportation sector (both terrestrial and air), 2 % of energy demand is met using biofuels, which is expected to rise to 3 % by 2020. The proportion of energy derived from biomass overall is not expected to change significantly over the coming decade (Canada National Energy Board 2011b).

Historically, the main source of feedstock for bioenergy production in Canada was mill residues that result from forest product manufacturing operations. However, the impact of lower mill residue production due to mill closures and new bioenergy projects outside the traditional forest sector left bioenergy plants scrambling for alternative biomass feedstock sources (Bradley 2010).

In Canada, the largest potential source of sustainable forest feed-stock for bioenergy production comes from the following: (1) harvest residues (i.e., tree tops and branches from forest harvesting operations), (2) salvageable trees from stands killed by natural disturbances (i.e., insects or wildfire) and (3) non-commercial tree species. In fact, harvest residues formed approximately 20–30 % of biomass feedstock in British Columbia pellet plants in 2010 (Bradley 2010), and also supplied biomass district heating systems across the province of Quebec.

Forest energy crops are being explored and now scaled-up at various locations across Canada. These plantations are positioned strategically to be accessible to final bioenergy users. However, high establishment and management costs relative to other forest biomass feedstock limit the appeal of this source (e.g. Yemshanov and McKenney 2008).

The harvested corn area in Canada has been relatively constant over the past 30 years but increases in agricultural productivity have led to an almost doubling of the total quantity of corn produced during that period. Ethanol production is now utilizing about 27 % of the Canadian corn crop. Since the corn is grown on land representing about 2.6 % of the managed agricultural land in Canada, less than 0.7 % of Canadian managed agricultural land is being used for corn for ethanol production. However, Canada is a net importer of corn (Canadian Renewable Fuels Association 2010).

Harvested wheat area in Canada has been declining due to more sustainable crop production methods. But the quantity of wheat produced has not dropped at the same rate as the area due to increases in crop yields. Less than 300,000 ha of Canada's managed agricultural land have been used on a gross basis for wheat-to-ethanol production (Canadian Renewable Fuels Association 2010).

In eastern Canada, the feedstock for biodiesel plants is usually animal fats. Alternative feedstock includes used vegetable oil and tallow. In western Canada (Manitoba to Alberta), the usual feedstock for biodiesel is oil derived from canola. Canola production has reached 11 million tons year^{-1} in recent years. Canada is a net ex-porter of canola and, therefore, no imports are required for canola feedstock (Canadian Renewable Fuels Association 2010)

4.4.2.2 Biomass for Energy – Production and Consumption

Production and export of wood pellets in Canada has grown significantly in the past several years, primarily on the west coast. Capacity grew from 500,000 tons in 2002 to around 2.93 Mt in 2011 (Wood Pellet Association of Canada 2012). For many pellet manufacturers, the primary fibre sources are now harvest debris and non-commercial round wood; in some cases these sources represent up to 70 % of feedstock, which bear a much higher cost than mill residues (Bradley 2010).

In 2004, around 12 % of the Canadian pellet production was for domestic consumption, whereas 36 % was exported to the United States and 52 % exported overseas, mainly to the United Kingdom and the Netherlands In 2010, the share of the Canadian production taken up by domestic consumption had fallen to 7 %, whereas the overseas market represented 90 % (Lamers et al. 2012). Since the domestic market is growing only marginally, it is expected that the majority of future production will be exported.

There were 59 cogeneration facilities operating at pulp and paper mills and sawmills across Canada in 2000, but this number dropped to 39 over the next decade after the closure of several pulp mills (Canadian Bioenergy Association 2012). These remaining cogeneration facilities have a total installed capacity of 1,349 MW of electrical output (MWel) and 5,331 MW of thermal output (MWth). The facilities use mill residues such as black liquor, bark/hog, shavings and sawdust as feedstock (Fig. 4.7).

The growth in the bioethanol production has been 150 % between 2004 and 2010 (compare Fig. 4.8). In 2010, Canada had 18 operating ethanol plants with a total

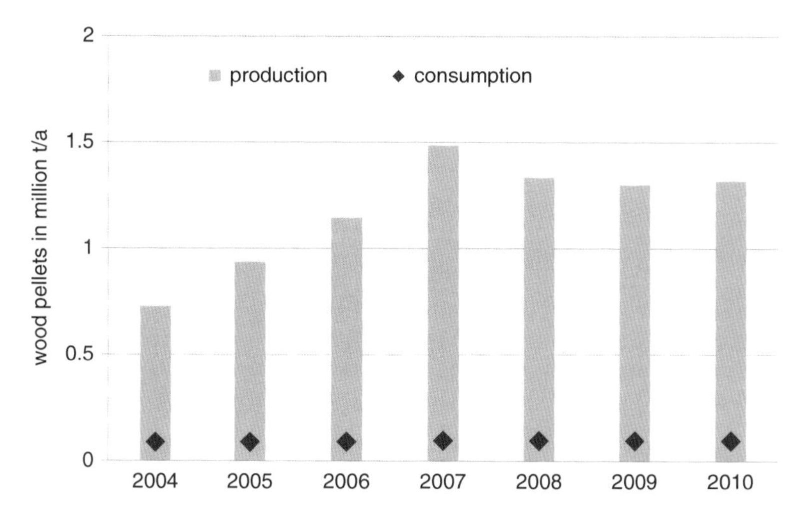

Fig. 4.7 Production and consumption of wood pellets in Canada, 2004–2010 (Lamers et al. 2012)

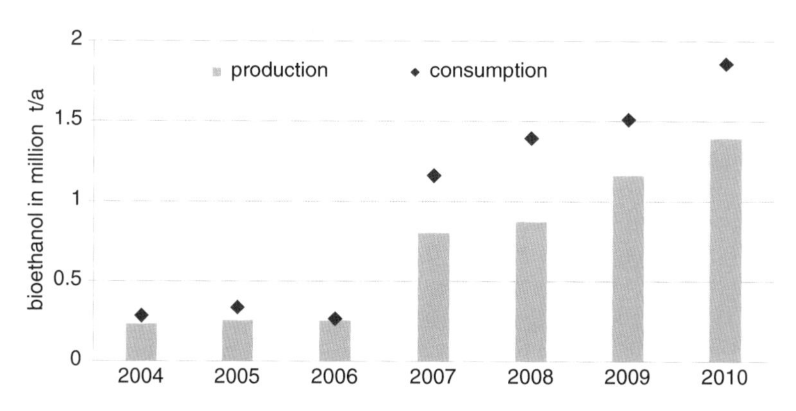

Fig. 4.8 Production and consumption of bioethanol in Canada, 2004–2010 (United States Energy Information and Administration 2012)

combined capacity of 1.8 billion litres/year (Canadian Renewable Fuels Association 2010). At full production capacity the ethanol facilities would create a demand for 93.0 million tons of corn, 1.4 million tons of wheat and 14,000 tons of cellulosic materials. Most plants in Western Canada use wheat as a feedstock, whereas corn is used in Eastern Canada. The province of Quebec prohibited any future corn ethanol plants, therefore in all likelihood all future ethanol plants in Quebec will be 2nd generation.

Growth in ethanol production capacity from 2004 to 2010 has been driven by the introduction of provincial and federal renewable fuel requirements (Canada Gazette 2010). Canada mandates an average 5 % renewable fuel content based on the gasoline volume and an average 2 % renewable fuel content in diesel fuel and heating distillate oil based on annual volumes. The renewable fuel mandates will

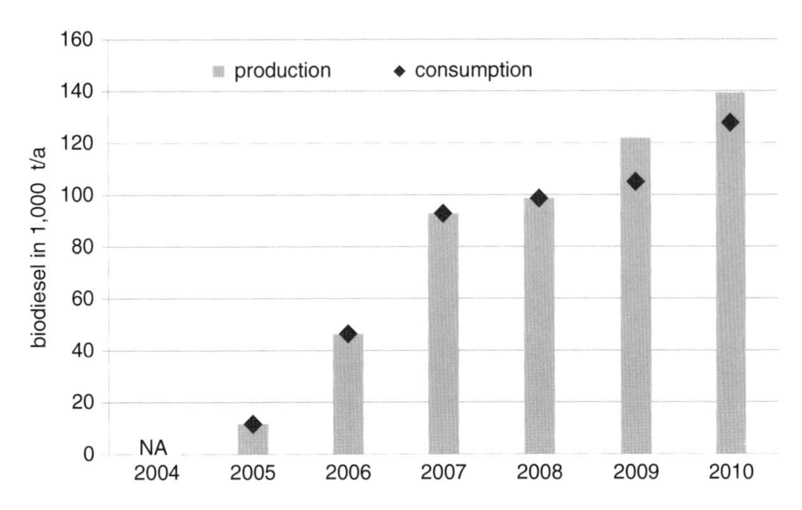

Fig. 4.9 Production and consumption of biodiesel in Canada, 2004–2010 (United States Energy Information and Administration 2012)

result in an ethanol demand of approximately 2 billion litres/year (Canadian Renewable Fuels Association 2010).

The Canadian biodiesel industry has not expanded as quickly as the ethanol industry, since the first provincially mandated market was only established at the end of 2009. Since 2004, the biodiesel production has grown from almost non-existent to 139 million litres/year in 2010 (compare Fig. 4.9). In 2010, Canada had 14 operating biodiesel plants with a total capacity of 206 million litres/year (Canadian Renewable Fuels Association 2010) Most of the biodiesel produced in Canada is from rendered animal fats and used cooking oils (also called yellow grease). Only a small quantity of biodiesel is produced from canola oil. However, some Canadian canola oil has been shipped to the United States to be transformed into biodiesel which is then shipped back to Canada to meet the demand in mandated markets (Canadian Renewable Fuels Association 2010).

It is estimated that there are about 400,000 tons of animal fats and waste oils produced per year in Canada. There is currently no domestic market for this material, and a large proportion of it is exported. The Canadian biodiesel industry may use up to 50 % of this production, depending on market demands for biodiesel (Canadian Renewable Fuels Association 2010).

4.4.2.3 Biomass for Energy – Prices

There is no official compilation of statistics on biomass prices in Canada, likely because a commodity market for biomass does not yet exist. However, evolution of the price of wood chips gives an indication of the potential market prices for biomass products. Trends in the delivered price of conifer wood chips, although largely used by the pulp & paper industry and to a smaller extent by the panel industry, give an indication of the minimum price bioenergy producers would have had to pay for this feedstock. Wood chip prices tend to vary by region with higher prices in eastern Canada relative to British Columbia. The price of softwood chips in Canada at the end of 2010 ranged from CAN$80-CAN$109 per oven-dry ton, while the price of hardwood chips was CAN$85 per oven-dry ton (RISI 2011). There is a slight downward trend since 2008 in North American wood chip prices, which is likely related to the troubles of the pulp & paper industry. If this trend is maintained, it may indicate that prices will in the short- to mid-term reach a level low enough to allow greater affordability as a feedstock for potential bioenergy producers.

In 2010, Canadian FOB industrial wood pellet prices in North American harbors ranged from CAN$130 to CAN$157 per oven-dry ton (excluding shipping costs) (Argus Media 2011). Most of the production of wood pellets for industrial consumption in western Canada is exported to Europe, while wood pellets produced in eastern Canada are typically sold domestically. Domestic North American wood pellet prices in 2010 ranged from CAN$195 to CAN$250 per oven-dry ton (Argus Media 2011).

4.4.2.4 Biomass for Energy – Trade

The current status of Canada's forest biomass market is largely one of export, and it is expected that the majority of future production will be exported. Europe is the main export market for Canadian pellets, but countries of the Pacific Rim, which are easily accessible for shipping from the west coast of Canada, are seen as the next important market, both for wood pellets and chips. Whereas most of pellet exports currently come from Western Canada, efforts are being made to tap into the biomass potential of the provinces of Eastern Canada (i.e. Ontario, Quebec, and the Maritimes) and increase their production, storage and shipping capacity. On the other hand, domestic markets are expected to grow only marginally at the national level. However, the growing support for cogeneration and community heating projects at the local, regional, provincial and federal levels, which is part of a push for the renewal of the forest industry and the revitalization of rural communities, will raise demand for solid biomass in some regions of Canada. More importantly, it will contribute to the establishment, economic stability and profitability of biomass supply chains that will also benefit Canada's international trade.

For liquid biofuels, while most Canadian production is produced and consumed domestically, there is some international trade. Canada has low barriers to trade compared with other OECD biofuel-producing countries, such as Australia, the United States and countries in the European Union (International Institute for Sustainable Development 2009). Until April 2008, imported ethanol was eligible for the same excise tax exemptions as domestically produced fuels, resulting in substantial imports from the United States and Brazil. The shift from excise tax exemptions to production subsidies in April 2008 changed the set. However, Canada's import tariff is zero on biofuels imported from the United States and other countries with which Canada has a free-trade agreement. The introduction of renewable fuel standards at the provincial and federal level also has an impact on biofuel trade in Canada. It is likely that imports of ethanol (notably from the United States, which is by far the largest source of imports) will continue in the near future and will help meet renewable fuel standards.

4.4.3 Brazil

4.4.3.1 General Introduction

Brazil is worldwide the fifth largest country by geographical area, the fifth most populous country (it surpassed 190 million people in 2010) and the 6th largest economy (surpassed the UK in 2011); and the largest Portuguese spoken country. It is located in South America and covers almost 50 % of the region; Brazil has boarders with all South American countries, except Chile and Ecuador. It is worth to note that in 2010, according to the World Bank (Trading Economics 2012), 61.4 % of the country's land area was still covered by forests.[4]

Brazil has one of the greenest energy mixes in the world. Biomass from sugar-cane and forestry residues used in transportation and electricity production accounted for 27.2 % of total energy consumption in 2010 and significantly contributed to the high of renewables in the Brazilian energy matrix.

Sugarcane is by far the most important source of biomass in Brazil and sugarcane ethanol is the benchmark and most important biofuel, especially after the introduction of the flex-fuel vehicles in 2003 which created the conditions for the rapid expansion of sugarcane between 2004 and 2010.

Although the main use of ethanol is in the transport sector, there is a growing interest for the product in the plastics and chemicals industries, with mills entirely dedicated to non-fuel markets (Fig. 4.10).

As a consequence of the rapid increase in the production of sugarcane derived products (including sugar), the use of sugarcane bagasse as a source for production of bioelectricity[5] doubled between 2004 and 2010 to 60 million tons. The share of renewables in the electricity production in Brazil was 87.2 % in 2010 (MME/EPE 2012).

Internal demand for fuel ethanol is expected to remain strong and reach 63.1 billion litres in 2020 (MME/EPE 2011). This expectation is based on the prediction of strong growth in the economy pushing for more flex-fuel cars in the market and the competitive prices of hydrous versus gasoline. The total share of biomass in the energy mix is expected to grow to 30.1 %.

[4] Defined as land area under natural or planted stands of trees of at least 5 m in situ, whether productive or not, and excludes tree stands in agriculture production systems.

[5] One of the interesting characteristics of the production of bioelectricity from sugarcane bagasse, making it even more attractive, is its complementary nature to hydropower, supplying electricity to the grid during the driest months between May and November.

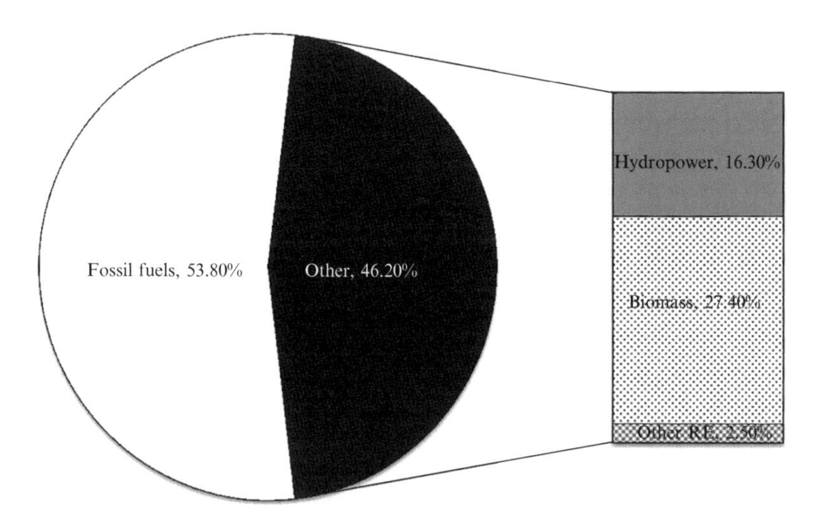

Fig. 4.10 Total final energy consumption in Brazil in 2010 (EPE 2011)

4.4.3.2 Biomass for Energy – Production and Consumption

Production of sugarcane ethanol has doubled between 2004 and 2010 to over 22 billion in 2010, pushed especially by growing internal demand with the introduction of the flex-fuel cars. In a 7 years period leading to 2010, the industry experienced a radical transformation, attracting large investments and going through a consolidation process that changed the industry, introducing new management practices and increasing transparency, consequently changing how it was perceived in the country but also abroad; it is considered as more opened and more professionalized now (Fig. 4.11).

In 2010, production of ethanol was high but consumption dropped due to climate conditions and as an effect of the 2008 crisis. Exports were especially low (1.9 billion litres), the lowest since 2004. Even with the classification of sugarcane ethanol as advanced biofuels in US and the introduction of E15 in that country, the exports to US reached only 500 million litres in 2010 (EPE/DPG 2011). The situation changed in 2012.[6]

In 2020, it is expected that ethanol will supply 50 % of the country's light automotive fleet and increase the share in the production of bioelectricity from sugarcane bagasse/leafs from 2 % in 2010 (1.1 GW) to 18 % in 2020 (15.3 GW).

The main challenges for the second half of this decade for the industry is to restore competitiveness of hydrous ethanol in the domestic market, guarantee a level playing field with the gasoline market and attract enough investments to finance new projects, growing operational costs (planting, storage, certification), increase efficiency and productivity (which remains high at 7,000 l/he). Specifically in the bioelectricity

[6] After importing 1.4 billion liters in 2011, exports of ethanol surged in 2012 to the highest volumes since 2009. Exports reached 3.04 billion liters (804 million gallons) in 2012 – the highest level since 2009 (MME 2013).

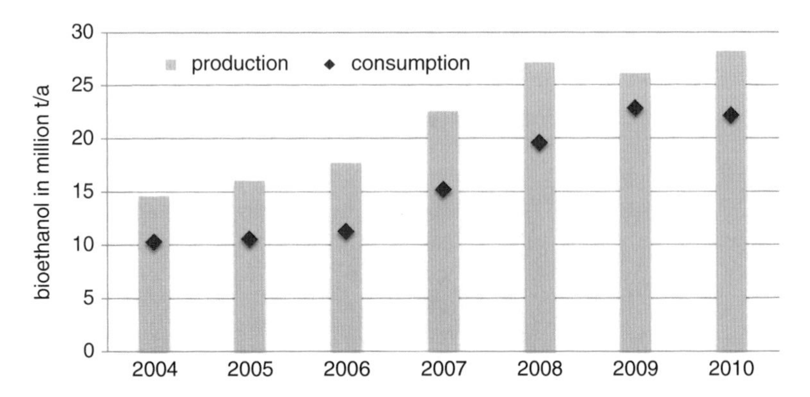

Fig. 4.11 Production and consumption of bioethanol in Brazil, 2004–2010 (ANP 2011)

area, the challenges include modernization of the industrial park and increasing productivity in order to sell more bioelectricity to the grid.

The Brazilian Biodiesel Program (PNPB) was launched in 2004 with the intention to create a biodiesel supply chain that could use different feedstock (castor beans, peanuts, cotton, palm oil, sunflower seeds and soybeans) from different regions of the country, promoting rural development and social inclusion.

It has innovated with the creation of the so called Social Fuel Label, a biofuels certification that mandates purchases from small family farmers in exchange for incentives such as eligibility to participate in official auctions and reduction in the payment of certain taxes. Certified producers in certain regions enjoy full tax exemption. The market is regulated by the Brazilian government through a public auction system which sets the volume of biodiesel that should be produced as well as the average sales price.

In 2010, 20 % of the feedstock was supplied by small family farmers (mostly in the south of the country) (UBRABIO/FGV 2010) and 34 biodiesel producers obtained this certification (MME 2011a). Also in 2010, the blending mandate of biodiesel was set to 5 %, which increased by 49 % the availability of B100 in the market. In only 4 years Brazil was able to produce 2.1 billion litres (Fig. 4.12) (EPE 2011).[7]

Currently, producers are allowed to sell only 60 % of their installed capacity. The reason for this cap is that total installed capacity has increased significantly to around 6 billion litres representing almost three times the obligation based on 5 % blending. The total volume of biodiesel traded in all auctions from 2005 to 2010 was 6.4 billion litres (EPE/DPG 2011).

The main challenge for the program is to obtain the diversification of feedstock. In 2010, soybean oil made up 82 % of the market followed by animal fat with 13.8 % (MME 2011b). A proposal to increase the mandate is under review by the Brazilian Government but it is expected that the current mandate will remain the

[7] 3.04 billion liters (804 million gallons) in 2012 – the highest level since 2009 (MME 2013).

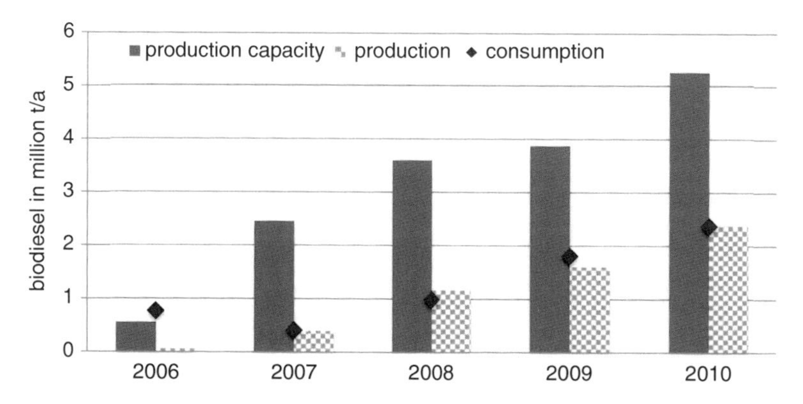

Fig. 4.12 Production capacity, production and consumption of biodiesel in Brazil, 2006–2010 (ANP 2005–2010)

same for coming years, especially if the goal for feedstock diversification is not reached. Nevertheless, the expectation is that, in 2020, production of Biodiesel will increase to 3.8 billion litres (EPE 2011).

In 2010, solid biomass from sugarcane bagasse and forestry/wood represented 21.1 % of total energy consumption (13.5 and 7.6 % respectively). In the electricity matrix, solid biomass had a share of 5.6 % in 2010 (EPE 2012a).

In order to incentivize production and use of biomass in the country, the Brazilian government via its development bank BNDES launched the Fundo Clima program to foster investments in local generation and distribution of renewable energy from biomass (excluding biomass from sugarcane which was incentivized via other instruments).

The increase in the production of sugarcane products (sugar and ethanol) between 2004 and 2010 led to the increase in the production of bagasse which grew 60 % to 160.3 million tons, from which 62 million was used in the production of electricity, doubled the volume used in 2004 and producing 1 GW in 2010, between 2 and 3 % of the Brazilian electricity mix. Still, the main destination of sugarcane bagasse was the industrial sector (83.6 million tons in 2010) (Fig. 4.13) (EPE 2012a).

The potential of sugarcane residues to be used as feedstock for the production of second generation biofuels has attracted numerous foreign companies to partner with Brazilian companies to develop the product. In 2014, it is expected that one of these partnerships will start production of ethanol based on sugarcane residues. Based in the North-east of the country, the company announced initial production capacity of 82 million litres/year and expects to produce 1 billion litres of ethanol in 2020. Other partnerships in the Center-South of Brazil have similar projects.

In 2010 consumption of forestry/wood was mostly for charcoal (33 %) followed by the industrial sector (27 %) and residential sector (28 %). By 2020, the combined consumption of biomass in Brazil (firewood and sugarcane bagasse) for energy purposes is expected to increase respectively by 21 % for firewood (57.3 million tons in 2010 to 69.4 million in 2020).

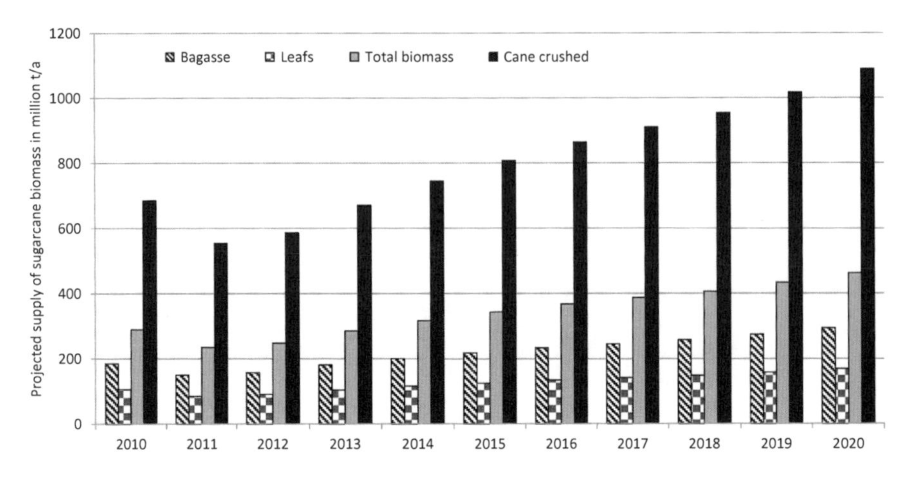

Fig. 4.13 Projected supply of sugarcane biomass until 2020 (EPE 2012a)

There is no specific information about production and use in Brazil of wood pellets. Recently, due to the crisis in Europe and slow-down in China, paper and cellulose companies are expanding their portfolios to include other forestry products such as wood pellets and that might trigger the first shipments of wood pellets to Europe in 2014. One of the challenges for the export of wood pellets from Brazil is the quality of port infrastructure to store and ship the product.

4.4.3.3 Biomass for Energy – Prices

Biodiesel prices are defined in public auctions where the National Petroleum Agency (ANP) sets the maximum prices. Producers are not allowed to change prices and therefore need to make sure they keep their production costs under control, especially the price of soybean oil which accounts for around 75 % of the cost of biodiesel production and consequently links the prices of biodiesel to the expectations of soybean oil prices (Fig. 4.14).

Historically, prices of soybean oil are linked to fossil fuel prices. If the expectation of high oil prices for this decade is confirmed then biodiesel prices in Brazil will follow suit. In 2010, the average price of 1 m^3 of biodiesel traded at official auctions was U$ 1.220 with 40 producers (34 of them certified) participating in the auction.

Ethanol Prices paid to producers in the state of São Paulo were at around U$ 0.6 per litre in 2010. Consumers saw a significant increase in prices in relation to 2009 and that has made the product less competitive with gasoline. The main reason was the low pace of growth of the supply and the decrease of demand still as a reflex of the 2008 crisis (Fig. 4.15).

Ethanol exports in 2010 (1.9 billion litres) were 58 % lower than 2009 and prices on average 22 % higher than in 2009, although prices in December were particularly similar (Fig. 4.16).

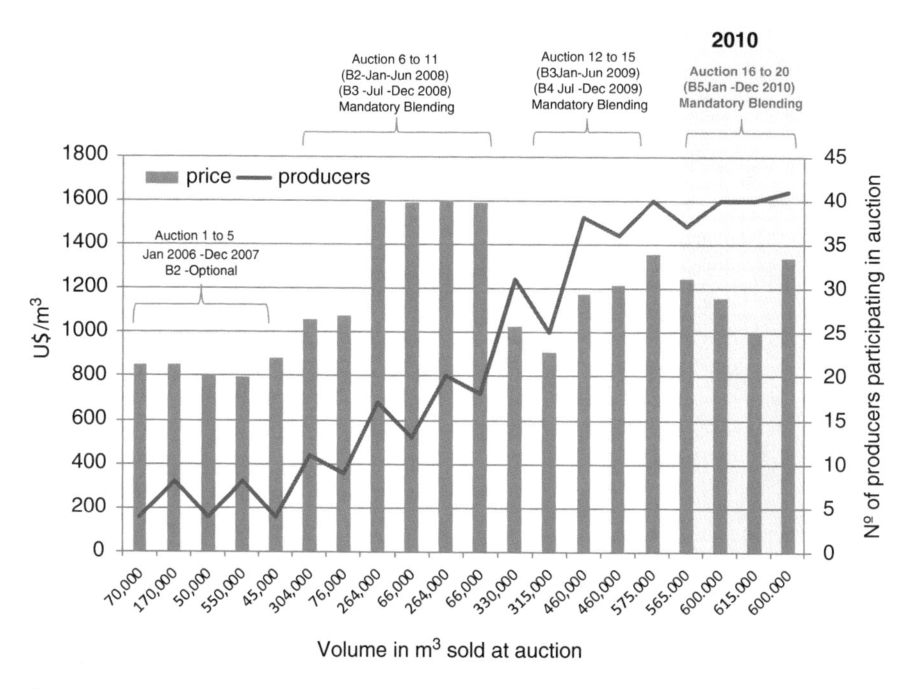

Fig. 4.14 Prices for biodiesel (ex-mill) – traded volumes, 2006–2010 (ANP 2005–2010)

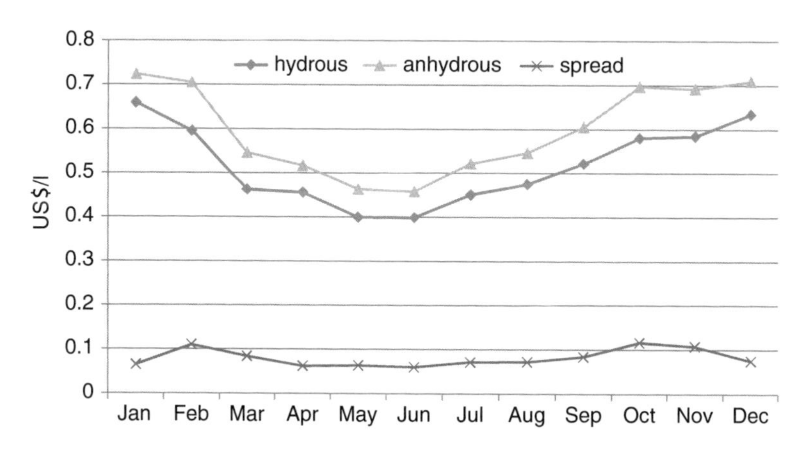

Fig. 4.15 Ethanol prices received by producers in São Paulo, 2010 (UNICADATA 2013)

4.4.3.4 Biomass for Energy – Trade

The question whether Brazil will regain the top position as ethanol exporter will depend on its capacity to improve efficiency and productivity and positive outcomes of the current regulatory framework that will shape trade with Europe and the US, the most important importers of ethanol from Brazil.

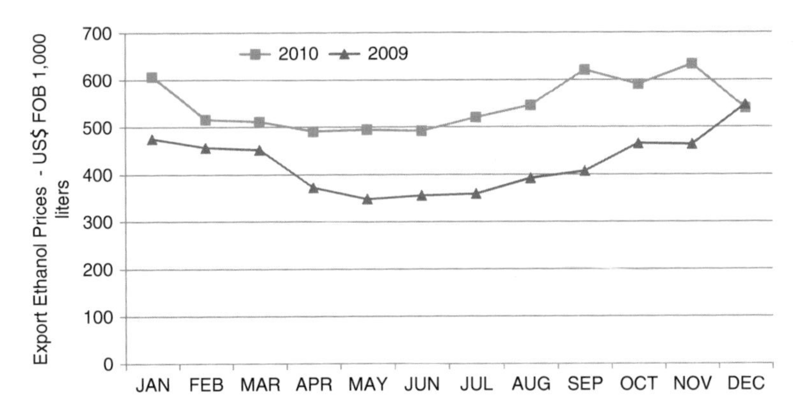

Fig. 4.16 Export prices for bioethanol, 2009–2010 (MDIC 2011)

In Europe, the new iLUC (indirect Land Use Change) proposal is introducing a cap in 5 % the use of food-crop/land-based biofuels, including sugarcane ethanol. It remains to be seen whether sugarcane ethanol, with its widely recognized sustainability credentials, often considered the best-in-class first generation biofuel, will be given a separate status from other food-crop based biofuels. The EU-27 will have a combined demand for ethanol of 14 billion litres in 2020.[8]

In the US, sugarcane ethanol is categorized as "non-cellulosic advanced biofuel" (with at least 50 % GHG reduction). The RFS2 program requirements for this category are expected to reach 13.2 billion liters in 2020 and the Brazilian ethanol might become the main supplier within this category (EPA 2012).

On the supply side, Brazil would be able to export between 3.3 and 14 billion litres in 2020 (MME/EPE and UNICA 2012). Exports already reached 3 billion in 2012.

Fuel ethanol exports to the US and Europe could reach 10.8 billion in 2020 if Brazil is able to cover 50 % of the non-cellulosic advanced biofuels market in US[9] and 30 % of the total ethanol demand in the EU-27. That would be more than double the historical record of 2008 when the country exported 5.1 billion litres. That would also mean Brazil would regain the top position as ethanol exporter.

4.4.4 Finland

4.4.4.1 General Introduction

Finland is a large and sparsely populated state: with a total area of 33.8 million ha, it is the fifth largest country in Europe and is located between 60° and 70° northern latitude. Finland has a population of 5.4 million, i.e. 17.5 people per square

[8] Potential demand for Ethanol in the EU in 2020: 14 billion liters (ECN 2011).

[9] Potential demand for Advanced Biofuels in the US in 2020: 13.5 billion liters.

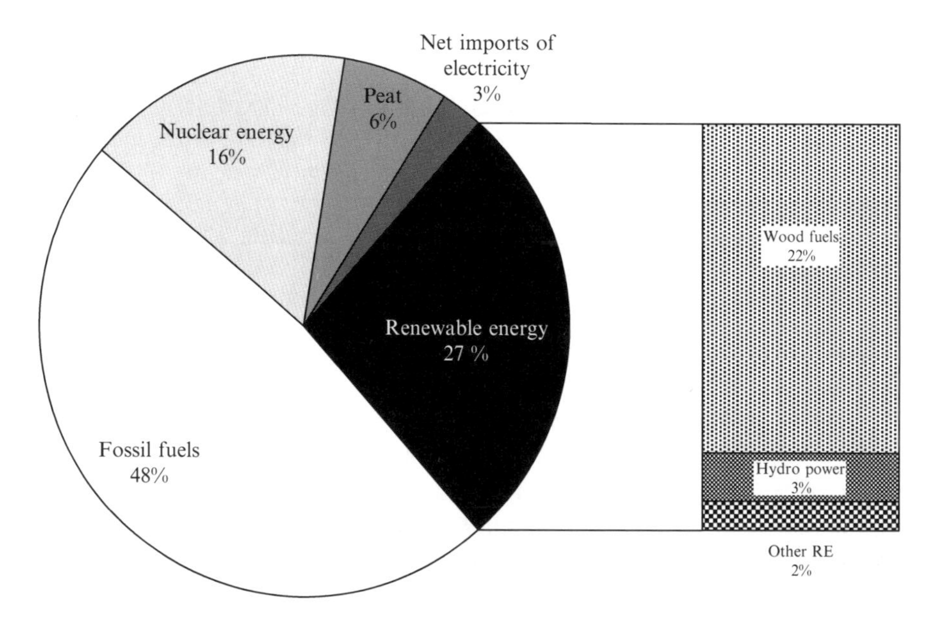

Fig. 4.17 Total final energy consumption in Finland, 2010 (The total use of primary energy in 2010 was 1,464 PJ) (Statistics Finland 2012)

kilometre.[10] Forestry land covers 87 % of the country's land area (30.4 million ha), only 9 % (2.8 million ha) is used for agriculture and the remaining 4 % consists of housing and urban development and transport routes.

Imported fossil fuels – oil, coal, and natural gas – have a major role as a primary energy source in the Finnish energy system, accounting for almost 50 % of the total primary energy supply (Fig. 4.17). The only significant indigenous energy resources in the country are wood, peat,[11] hydropower, and wind energy. In 2010, renewable energy sources accounted for 27 % (Statistics Finland 2012).

In Finland, primary energy consumption per capita is high, 287 MJ/capita in 2008 (Statistics Finland 2011a). For comparison, in the same year the corresponding figure for the EU-27 countries was 151 MJ/capita (Statistics Finland 2011a). The cold climate, long distances, high standard of living and energy intensive structure of industry are factors that result in high specific energy consumption. In Finland, industry consumes nearly half of all energy. The forest industry represents the largest producer of wood fuels, but the industry is also a major user of wood fuels. Almost two thirds of wood fuels use takes place in the forest industry. Wood is the most important fuel at forest industry mills, accounting for about 75 % of their fuel consumption (Peltola 2008). In many cases, paper, paperboard, pulp and saw mills are

[10] Population density counted for land area.

[11] In Finland, peat has been defined as a slowly renewing biomass fuel (Ministry of Employment and the Economy of Finland 2009). It is not considered a renewable energy source in official statistics and in greenhouse gas accounting.

located on the same site, forming a forest industry integrate which allows efficient utilization of raw material and energy.

The RES Directive of the IEU has set 38 % as a target for the share of renewable energy in final energy consumption in Finland in 2020 (European Commission 2009). In 2010, the realized share of renewable energy in the final energy consumption was estimated to be nearly 30 %.

4.4.4.2 Biomass for Energy – Production and Consumption

Forest biomass is the most important source of renewable energy in Finland, covering approximately 80 % of the renewable energy used. Most forest-based bioenergy (over 75 %) is generated from by-products of the forest industry (black liquor, bark, and sawdust). The rest of the wood energy is generated from wood biomass that is sourced from forests for energy purposes (firewood and forest chips). The proportion of wood pellets has been negligible. The volumes of the forest industry's energy by-products vary with the production of pulp and paper.

In the current energy policy, (Ministry of Employment and the Economy 2008, 2010; Pekkarinen 2010) the target for the use of forest chips (logging residues, stumps, and energy wood) in 2020 has been set at 13.5 million solid cubic meters (equalling approximately 97 PJ). Forest chips will account for most future growth in renewable energy production. Forest chips are expected to become an important raw material in the production of liquid biofuels.

In recent years, a significant capacity for production of biofuels has been constructed in Finland, covering bio-ETBE,[12] hydro-treated *biodiesel* (NExBTL), and *bioethanol*. The production of ETBE and NExBTL is mostly based on imported raw materials (bioethanol and palm oil). Finnish bioethanol is produced from non-cellulosic raw materials such as by-products and waste streams of the food industry. Compared to the current use of transport biofuels in Finland, the existing production capacity is almost 20 PJ/year. However, the existing production capacity will not meet the target level set for biofuels in road transport in 2020, and either more domestic capacity is required or the biofuel import has to be increased.

Wood pellet production in Finland started in the end of 1990s. The Finnish pellet industry was founded on export supplying pellets to Sweden, where pellet markets were developing rapidly at the time. The majority of Finnish pellet production has been consumed abroad (Fig. 4.18). The number of export countries of pellets has increased resulting from booming pellet markets in Europe. In addition to Sweden, Finnish pellets have been exported to Denmark, the Netherlands, the UK, and Belgium. At the beginning of 2011, there were 28 wood pellet mills in operation. The total production capacity of the pellet mills is approximately 700,000 tons/year.

---

[12] ETBE (ethyl-tertio-butyl-ether) is an additive that enhances the octane rating of petrol (replacing lead and benzene in unleaded petrol) and reduces emissions. Bio-ETBE is produced by combining bioethanol and fossil isobutylene.

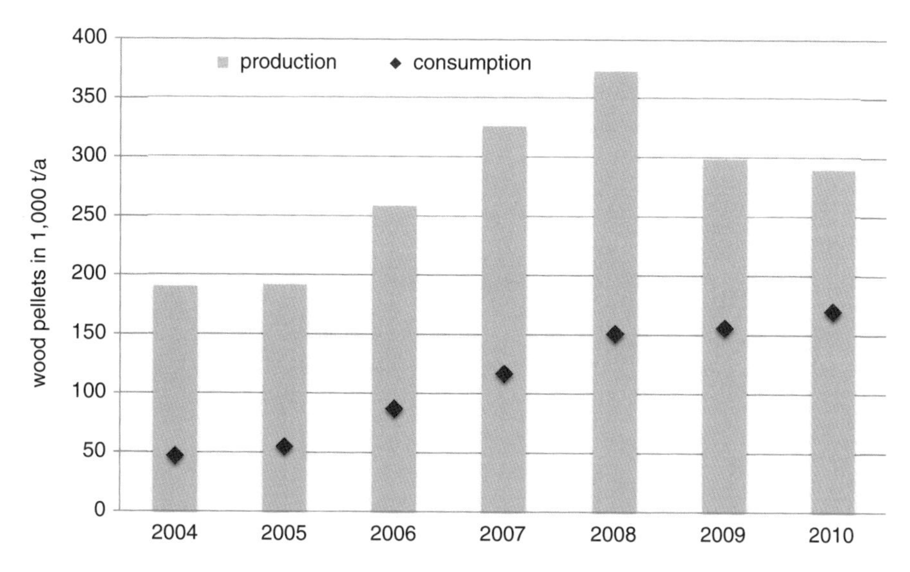

Fig. 4.18 Production and consumption of wood pellets in Finland, 2004–2010 (Statistics Finland 2012)

4.4.4.3 Biomass for Energy – Prices

In Finland, fuels used in the production of electricity are exempt from energy taxes, whereas in heat production, taxes are levied on some fuels. In heat production, fossil fuels and tall oil are taxed and the total prices of the fuels consist of market prices and taxes. The energy taxation of fossil fuels changes the mutual competitiveness of the fuels based on market prices. The energy taxation has rendered the consumer prices of heating oils and coal higher compared to wood chips from forests. Wood pellets are less expensive than light fuel oil, but are not competitive against heavy fuel oil, natural gas and coal in heat production (Fig. 4.19). In a longer 15-year period, the price development of indigenous fuels (wood and peat) has been moderate and stable compared to prices of fossil fuels, which have fluctuated remarkably mainly due to world market prices.

In Finland, woody by-products from the forest industry are fully utilized as raw material or in energy production, and their use cannot be increased unless the production volumes of the forest industry increase. Logging residues, stumps and small-diameter energy wood constitute a large underutilized biomass fuel potential. Increasing the use of forest chips in heating and power plants has an important role in the Finnish energy policy in decreasing CO_2 emissions from energy production. In Finland, the use of forest chips in heat and power plants has been increasing moderately since the 1980s. The increased consumption of forest-based fuels and strong development of technologies for forest chips production within national technology programs have lowered the prices of forest fuels during the 1990s. Since the turn of the millennium, the prices of forest chips have increased (Fig. 4.20). The measures of the domestic energy policy have boosted

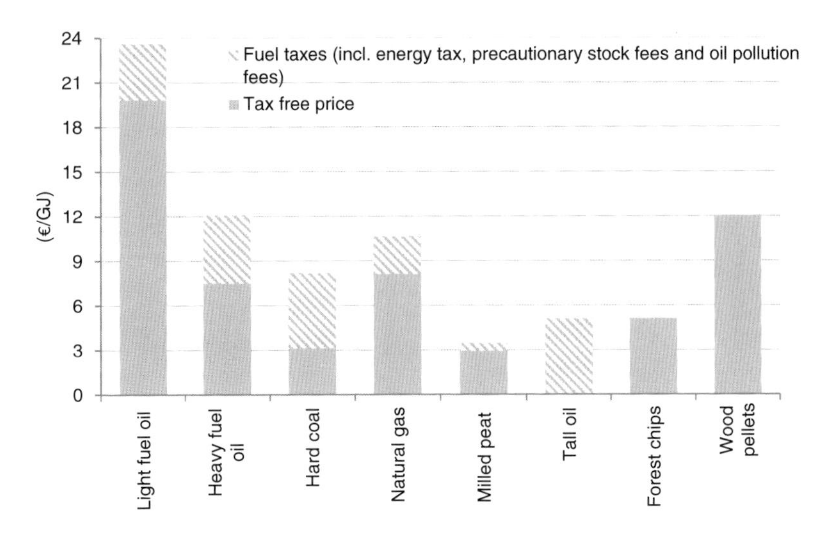

Fig. 4.19 Composition of fuel prices in heat production in Finland, January 2011 (Finnish Government 2010a, b; Statistics Finland 2011b). Value added tax (VAT) 23 % excluded. VAT is added for private consumers' prices. The price of tall oil was not available

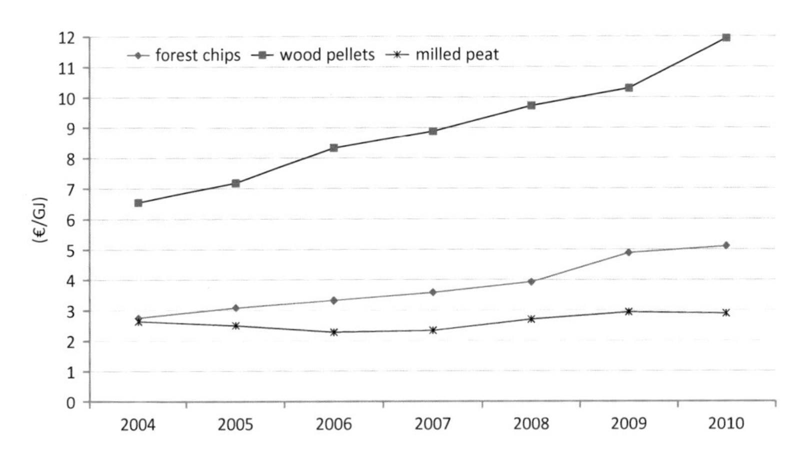

Fig. 4.20 Prices for forest chips, milled peat and wood pellet (delivered to plants), 2004–2010. Value Added Tax (VAT) excluded (Sources: 2000–2006 (Ylitalo 2001, 2002, 2003, 2004, 2005, 2006, 2007[13]))

the demand for biomass fuels, which has caused an upward trend in prices of wood fuels in recent years. Since the beginning of 2005, the start of the trading of CO_2 emission allowances within the EU emission trading scheme has enhanced the paying capacity of power plants for biomass fuels, and forest fuels have been to a greater extent produced at sites where the production costs are higher and the

[13]Ylitalo E, Finnish Forest Research Institute. Personal communication. E-mail: 31 May 2007.

production has previously been uneconomical. Also the utilization of costlier raw materials, small-diameter wood and stumps in addition to logging residues has increased in the production of forest chips.

4.4.4.4 Biomass for Energy – Trade

Finland is a large net importer of biomass fuels. Most of the imports are indirect and take place within the forest industry's raw wood imports. Wood pellets and tall oil form the majority of export streams of biomass fuels. The international trade of biomass fuels has a substantial importance for the utilization of bioenergy in Finland. In 2009, the total international trading of solid and liquid biomass fuels was approximately 45 PJ, of which import was 23 PJ. The indirect import of wood fuels which takes place within the forest industry's raw wood import grew until 2006, being 61 PJ in that year. In 2009, the import of raw wood collapsed, and correspondingly, the indirect import of wood fuels dropped to 23 PJ. In 2004–2008, wood pellets and tall oil formed the majority of export streams of biomass fuels. During 2007–2009, two large biodiesel production units, with the capacity of 380,000 tons/year in total, were established in Porvoo, and palm oil and biodiesel have become the largest import and export streams of energy biomass (Heinimö and Alakangas 2011b).

4.5 Summary and Conclusion

Biomass is the most relevant renewable energy carrier in many countries. It is used for the provision of heat, electricity and biofuels for the transport sector. The use in traditional stoves, existing fossil fuel conversion systems and modern, dedicated biomass conversion plants are prominent applications. The majority of the OECD countries sets ambitious targets for the next decade and has implemented different support schemes primarily for liquid biofuels and power generation from biomass.

Nevertheless the relevance of solid, liquid and gaseous biofuels differs between the countries, as well as the typical concepts and capacities of production and utilization plants and the support schemes. This might be caused by various factors. The case studies have revealed that the biomass potential situation and the targets of the energy policy play a major role.

In most of the cases the market development starts with a consideration of the domestic resources. Data availability on the resource situation is one precondition for a clear target definition on the development of the bioenergy sector. However, this information is often difficult to obtain from the statistics. Also shared responsibilities of national and federal governance concerning energy policy can be an impediment for the provision of biomass potential information. Thus cooperation between federal and national governmental bodies and between the responsible authorities for forestry, agriculture, waste management and energy is needed for the development of the market.

Another clear outcome of the case studies is that the development of bioenergy production and use strongly depends on the political support schemes. Especially for the development of commodities for trade like pellets, biodiesel, bioethanol (and biomethane in the future), trustable, long-term and stable political framework conditions are needed. The case studies showed that in Brazil, Finland and Germany those conditions were given for certain sectors during the last decade and gradually created consolidated markets, i.e. for bioethanol in Brazil, for heat and power production from woody biomass in Finland and for biodiesel and biogas in Germany. This progress can also be seen as a basis for the development of bioenergy trade.

The role of trade in the different countries also depends on the specific resource situation, taking into account the overall biomass potential and the expected domestic demand. Hence, for some countries resource limitations can be expected to become relevant when implementing national targets for bioenergy. Consequently, ambitious targets in almost all OECD countries might lead to a comparable higher import demand.

Trade just based on market prices for fossil fuels and bioenergy carriers is currently of minor relevance. This will change if the prices for fossil fuels and/or for carbon emission certificates increase; both is expected to happen in the long run, but difficult to predict.

References

AGEB e.V. (2011). *Auswertungstabellen zur Energiebilanz Deutschland*. Daten für die Jahre von 1990 bis 2010.

Agriculture and Agri-Food Canada. (2012). *An overview of the Canadian agriculture and agri-food system. Report*. Ottawa: Government of Canada, 175 p.

ANP. (2005–2010). *Estatísticas Biocombustíveis*. Resumo dos leilões de biodiesel da ANP.

ANP. (2011). *Production of ethanol 2001–2010 and ethanol sales 2000–2010*. Anuário Estatístsico 2011. Available at: http://www.anp.gov.br/?pg=56346#Se__o_4

Argus Media. (2011). *Biomass markets*, January–December 2010.

BAFA. (2011). *Amtliche Mineralöldaten*. Biodieselproduktion und -verbrauch. 2004–2010.

Bemelmans-Videc, M., et al. (2010). *Carrots sticks & sermons. Policy instruments & their evaluation*. New Brunswick: Transaction Publishers.

BMU. (2011). *Erneuerbare Energien 2010*. Daten des Bundesministeriums für Umwelt, Naturschutz und Reaktorsicherheit zur Entwicklung der erneuerbaren Energien in Deutschland im Jahr 2010 auf der Grundlage der Angaben der Arbeitsgruppe Erneuerbare Energien-Statistik (AGEE-Stat).

BMU. (2013). *EEG-dialog*. Gesprächsreihe zur EEG-Reform. Available at: http://www.bmu.de/themen/klima-energie/energiewende/eeg-reform

Bradley, D. (2010). *Canada report on bioenergy 2010* Submitted to the International Energy Agency – Bioenergy Task 40 Report. Ottawa: Canada, 53 p.

Bundesverband der deutschen Bioethanolwirtschaft e.V. (BDBE e.V.). (2009). *Preisentwicklung von Bioethanol*.

Canada Gazette. (2010). Renewable fuels regulations. *Environment Canada, 144*(15), April 10. Available at: http://www.gazette.gc.ca/rp-pr/p1/2010/2010-04-10/html/reg1-eng.html

Canada National Energy Board. (2011a). *Canadian energy overview 2010. Energy briefing note*. Ottawa: Government of Canada. 24 p.
Canada National Energy Board. (2011b). *Canada's energy future energy supply and demand projections to 2035*. Ottawa: Government of Canada. 79 p.
Canadian Bioenergy Association. (2012). *Economic impact of bioenergy in Canada 2011. Report*. Ottawa, 26 p.
Canadian Renewable Fuels Association. (2010). *Growing beyond oil: Delivering our energy future. A report card on the Canadian renewable fuels industry*. Ottawa, 50 p.
Centrales Agrar-Rohstoff-Marketing- und Entwicklungs-Netzwerk e.V. (C.A.R.M.E.N. e.V.). (2011a). *Preisentwicklung bei Holzpellets – Der Pellet-Preis-Index*. Available at: http://www.carmenev.de/dt/energie/pellets/pelletpreise.html
Centrales Agrar-Rohstoff-Marketing- und Entwicklungs-Netzwerk e.V. (C.A.R.M.E.N. e.V.). (2011b). *Preisentwicklung bei Waldhackschnitzeln – der Energieholz-Index*. Available at: http://www.carmen-ev.de/dt/energie/hackschnitzel/hackschnitzelpreis.html
Cocchi, M. (2012). *Country profile Italy 2011* (IEA bioenergy task 40 sustainable international bioenergy trade). Available at: http://www.bioenergytrade.org/downloads/iea-task-40-country-report-2011-italy.pdf
Cocchi, M., Marchal, D., Nikolaisen, L., Junginger, H. M., Goh, C. S., Heinimö, J., Bradley, D., Hess, R., Jacobson, J., Ovard, L., Thrän, D., Hennig, C., Deutmeyer, M., & Schouwenberg, P. P. (2011). *Global wood pellet industry market and trade study* (IEA bioenergy task 40 sustainable international bioenergy trade). Available at: http://bioenergytrade.org/downloads/t40-global-wood-pellet-market-study_final.pdf
DBFZ. (2011). *Global and regional spatial distribution of biomass potentials – status quo and options for specification* (DBFZ Report No. 7). Leipzig.
DBFZ. (2012a). *Monitoring zur Wirkung des Erneuerbare-Energien-Gesetz (EEG) auf die Entwicklung der Stromerzeugung aus Biomasse* (DBFZ Report No. 12). Leipzig.
DBFZ. (2012b). *Monitoring Biokraftstoffsektor* (DBFZ Report No. 11). Leipzig.
Deutscher Energieholz- und Pellet-Verband e.V. (DEPV e.V.). (2011a). *Entwicklung Pelletproduktion in Deutschland*. Available at: http://www.depv.de/startseite/marktdaten/entwicklungpelletproduktion/
Deutscher Energieholz- und Pellet-Verband e.V. (DEPV e.V.). (2011b). *Preisentwicklung Pellets*. Available at: http://www.depv.de/startseite/marktdaten/pelletspreise
ECN. (2011). *Renewable Energy Projections* as published in the National Renewable Energy Action Plans of the European Member States. Covering all 27 EU Member States with updates for 20 Member States, 28 November 2011.
EPA. (2012). *Renewable Fuel Standard program (RFS)*.
EPE. (2011). *Balanço Energético Nacional 2011: Ano Base 2010*.
EPE. (2012a). *PDE – Plano Decenal de Expansão de Energia 2020*.
EPE. (2012b). *Balanço Energético Nacional 2012: Ano base 2011*. Empresa de Pesquisa Energética.
EPE/DPG. (2011). *Análise da Conjuntura dos Biocombustíveis Jan 2010–Dec 2010*.
European Commission. (2009). *Directive2009/28/EC of the European parliament and of the council of 23 April 2009 on the promotion of the use of energy from renewable sources and amending and subsequently repealing Directives 2001/77/EC and 2003/30/EC*. Brussels, Belgium, 47 p.
Federal Republic of Germany. (2010). *National Renewable Energy Action Plan in accordance with Directive 2009/28/EC on the promotion of the use of energy from renewable sources*. Berlin, Germany.
Finnish Government. (2010a). *Laki nestemäisten polttoaineiden valmisteverosta* 29 December /1472.
Finnish Government. (2010b). *Laki sähkön ja eräiden polttoaineiden valmisteverosta* 30 December 1996/1260.
FNR. (2013). *Bioenergy in Germany*. Available at: http://www.biodeutschland.org/tl_files/content/dokumente/biothek/Bioenergy_in-Germany_2012_fnr.pdf
F. O. Licht's (2011). World ethanol & biofuels report Bd. 2008–2011. London, United Kingdom.
Fraunhofer UMSICHT. (2012). *Overview of biomethane markets and regulations in partner countries*. EU IEE project Green Gas Grids. Oberhausen, Germany.

Goh, C. S., Junginger, M., & Faaij, A. (2012). *Country report for the Netherlands 2011* (IEA bioenergy task 40). Utrecht, Netherlands.
Guisson, R., & Marchal, D. (2011). *Country report Belgium 2011* (IEA bioenergy task 40).
Heinimö, J., & Alakangas, E. (2011a). *Country report of Finland 2011* (IEA bioenergy task 40).
Heinimö, J., & Alakangas, E. (2011b). *Market of biomass fuels in Finland – An overview 2009* (IEA Bioenergy Task 40 and Eubionet III). Country report of Finland, Lappeenranta: Lappeenranta University of Technology. Lappeenranta, Finland, 40 p.
Hektor, B. (2012). *Country report Sweden 2011* (IEA bioenergy task 40). Stockholm, Sweden.
IEA. (2006a). *Energy balances of non-OECD countries 2003–2004.*
IEA. (2006b). *Energy balances of OECD countries 2003–2004.*
IEA. (2011). *World Energy Outlook 2011.*
International Institute for Sustainable Development. (2009). *Biofuels: At what cost? Government support for ethanol and biodiesel in Canada. Report for the Global Subsidies Initiative.* Winnipeg/Geneva, 13 p.
Kalt, G., et al. (2011). *Country report: Austria 2011* (IEA bioenergy task 40). Vienna, Austria.
Lamers, P., Junginger, M., Hamelinck, C., & Faaij, A. (2012). Developments in international solid biofuel trade – An analysis of volumes, policies, and market factors. *Renewable and Sustainable Energy Reviews, 16*, 3176–3199. doi:10.1016/j.rser.2012.02.027.
MDIC. (2011). *EXPORTAÇÃO – PRODUTOS SELECIONADOS*. Available at: http://www.desenvolvimento.gov.br/arquivos/dwnl_1294316538.zip
Ministry of Employment and the Economy. (2008). *Pitkän aikavälin ilmasto- ja energiastrategia* [Long-term Climate and Energy Strategy]. Helsinki, 130 p.
Ministry of Employment and the Economy. (2010). *Suomen kansallinen toimintasuunnitlema uusiutuvista lähteistä peräisin olevan ennergian edistämisestä direktiivin 2009/28/EY mukaisesti* (NREAP Finland), 19 p.
Ministry of Employment and the Economy of Finland. (2009) *Uusiutuvat energialähteet ja turve*. Retrieved June 17. Available at: http://www.tem.fi/index.phtml?s=2481
MME. (2011a). *Boletim dos Combustíveis Renováveis*. Edição 42.
MME. (2011b, February). *Boletim Mensal dos Combustíveis Renováveis*. Edição nº 38.
MME. (2013). *Boletim Mensal dos Combustíveis Renováveis*, ed. nº 60.
MME/EPE. (2011). *Plano Decenal de Expansão de Energia 2020.* Ministério de Minas e Energia. Empresa de Pesquisa Energética.
MME/EPE. (2012). *Balanço Energético Nacional.*
MME/EPE, & UNICA. (2012). *Plano Decenal de Expansão de Energia 2021.*
Natural Resources Canada. (2011). *The state of Canada's forests – Annual report 2011*. Ottawa: Government of Canada, 52 p.
Natural Resources Canada. (2012). *The state of Canada's forests – Annual report 2012*. Ottawa: Government of Canada, 56 p.
Nikolaisen, L. (2012). *Country report 2011 for Denmark* (IEA bioenergy task 40). Arhus, Denmark.
Pekkarinen, M. (2010, April 20). *Kohti vähäpäästöistä Suomea – Uusiutuvan energian velvoitepaketti*. Power Point presentation. . Helsinki: Ministry of Employment and the Economy.
Peltola, A. (2008). *Metsätilastollinen vuosikirja 2008* [The Finnish statistical yearbook of forestry 2008]. (ed.), Finnish Forest Research Institute. Vantaa, Finland, 456 p.
RISI. (2011, April). *Wood biomass market report*.
Rosillo-Calle, F., & Galligani, S. (2011). *Country report for United Kingdom 2011* (IEA Bioenergy Task 40). London, United Kingdom.
Solar Promotion GmbH. (2010). Pelletsproduktion in Deutschland. *Pellets Markt und Trends,* Issue, 06–10.
Statistics Canada. (2012). *Census of agriculture 2011*. Available at: http://www.statcan.gc.ca/pub/95-640-x/2012002-eng.htm. Accessed Oct 2012.
Statistics Finland. (2011a). *Energy statistics. Yearbook 2010*. Helsinki: Official Statistics of Finland.

Statistics Finland. (2011b). *Energy prices*. Available at: http://pxweb2.stat.fi/database/statfin/ene/ehi/ehi_en.asp. Accessed 10 Nov 2011.

Statistics Finland. (2012). *Energy statistics. Yearbook 2011*. Helsinki: Official Statistics of Finland.

Thrän, D. et al. (2012). *Country report Germany 2011* (IEA bioenergy task 40). Leipzig, Germany.

Trading Economics. (2012). *World Bank indicators – Brazil – Land use*. Available at: http://www.tradingeconomics.com/brazil/forest-area-percent-of-land-area-wb-data.html

Tromborg, E. (2011). *Country report for Norway 2011* (IEA bioenergy task 40). As, Norway.

UBRABIO/FGV. (2010). *O biodiesel e sua contribuição ao desenvolvimento brasileiro*.

Unicadata. (2013). *Preço ao Produtor in São Paulo*. Available at: http://www.unica.com.br/unicadata

United States Energy Information and Administration. (2012). *International energy statistics*. Page accessed Sept 2012 (online). Available at: http://www.eia.gov/cfapps/ipdbproject/IEDIndex3.cfm

VDB. (2011). *Marktdaten*. Available at: www.biokraftstoffverband.de

Walter, A., & Dolzan, P. (2012). *Country report Brazil 2011* (IEA bioenergy task 40). Campinas, Brazil.

Wood Pellet Association of Canada. (2012, October 9–10). *Canada's wood pellet industry, status Quo and outlook*. Presentation at 12th Pellet Industry Forum. Berlin, Germany. Available at: http://www.pellet.org/images/2012-10-10_G_Murray_Cdn_Event.pdf

Yemshanov, D., & McKenney, D. (2008). Fast-growing poplar plantations as a bioenergy supply source for Canada. *Biomass and Bioenergy, 32*, 185–197.

Ylitalo, E. (2001). *Puupolttoaineen käyttö energiantuotannossa vuonna 2000*. Metsätilastotiedote 574. 4.5.2001. Official Statistics of Finland (Ed.). Helsinki, Metsäntutkimuslaitos, 5 p.

Ylitalo, E. (2002). *Puupolttoaineen käyttö energiantuotannossa vuonna 2001*. Metsätilastotiedote 620. 22.4.2002. Official Statistics of Finland (Ed.). Helsinki, Metsäntutkimuslaitos, 5 p.

Ylitalo, E. (2003). *Puupolttoaineen käyttö energiantuotannossa vuonna 2002*. Metsätilastotiedote 670. 25.4.2003. Official Statistics of Finland (Ed.). Helsinki, Metsäntutkimuslaitos, 5 p.

Ylitalo, E. (2004). *Puupolttoaineiden käyttö energiantuotannossa 2003*. Metsätilastotiedote 719. 6.5.2004. Official Statistics of Finland (Ed.). Helsinki, Metsäntutkimuslaitos, 7 p.

Ylitalo, E. (2005). *Puupolttoaineiden käyttö energiantuotannossa 2004*. Metsätilastotiedote 770. Official Statistics of Finland (Ed.). Helsinki, Metsäntutkimuslaitos, 7 p.

Ylitalo, E. (2006). *Puupolttoaineiden käyttö energiantuotannossa 2005*. Metsätilastotiedote 820. 4.5.2006. Official Statistics of Finland (Ed.). Helsinki, 7 p.

Chapter 5
Optimization of Biomass Transport and Logistics

Erin Searcy, J. Richard Hess, JayaShankar Tumuluru, Leslie Ovard, David J. Muth, Erik Trømborg, Michael Wild, Michael Deutmeyer, Lars Nikolaisen, Tapio Ranta, and Ric Hoefnagels

Abstract Global demand for lignocellulosic biomass is growing, driven by a desire to increase the contribution of renewable energy to the world energy mix. A barrier to the expansion of this industry is that biomass is not always geographically where it needs to be, nor does it have the characteristics required for efficient handling, storage, and conversion, due to low energy density compared to fossil fuels. Technologies exist that can create a more standardized feedstock for conversion processes and decrease handling and transport costs; however, the cost associated with those operations often results in a feedstock that is too expensive. The disconnect between quantity of feedstock

---

E. Searcy • J. Tumuluru • L. Ovard
Biofuels and Renewable Energy Technologies, Idaho National Laboratory, Idaho Falls, ID, USA
e-mail: erin.searcy@inl.gov; jayashankar.tumuluru@inl.gov; leslie.ovard@inl.gov

J.R. Hess (✉)
Division Director, Energy Systems and Technologies, Idaho National Laboratory, 1625, Idaho Falls, Idaho
e-mail: jrichard.hess@inl.gov

D.J. Muth
Praxik, LLC
e-mail: david.muth@praxik.com

E. Trømborg
Department of Ecology and Natural Resource Management, Norwegian University of Life Sciences, Ås, Norway
e-mail: erik.tromborg@umb.no

M. Wild
Wild & Partner KG, Vienna, Austria
e-mail: michael@wild.or.at

M. Deutmeyer
Biomass & Bioenergy, Green Resources AS, London, UK
e-mail: Michael.Deutmeyer@green-carbon-group.com

M. Junginger et al. (eds.), *International Bioenergy Trade: History, status & outlook on securing sustainable bioenergy supply, demand and markets*, Lecture Notes in Energy 17, DOI 10.1007/978-94-007-6982-3_5,

needed to meet bioenergy production goals, the quality required by the conversion processes, and the cost bioenergy producers are able to pay creates a need for new and improved technologies that potentially remove barriers associated with biomass use.

Because of their impact on feedstock cost, feedstock location and raw physical format are key barriers to industry expansion and intercontinental trade. One approach to reducing biomass cost is to emulate the commodity fossil-fuel-based feedstocks that biomass must compete with in terms of logistics, quality, and market characteristics. This requires preprocessing the biomass to improve density, flowability, stability, consistency, and conversion performance. Making the biomass format compatible with existing high-capacity transportation and handling infrastructure will reduce the need for new infrastructure. Producing biomass with these characteristics at costs conducive to energy production requires the development of new technologies or improvements to existing ones.

5.1 The Role of Transport and Logistics in Achieving Global Bioenergy Targets

The volume of biomass required to support a bioenergy industry that is capable of realizing even just a tenth[1] of the world's bioenergy production potential is often overlooked. At full capacity, sustainable global bioenergy production potential has been estimated sufficient to offset up to 60 % of world primary energy demand (Berndes et al. 2003; Smeets et al. 2007; Campbell et al. 2008; IEA 2008), and is technically capable of providing up to 1,500 EJ/year by 2050 (Bauen et al. 2009).

To realize substantive environmental benefits of bioenergy production, such as greenhouse gas reduction of 50 % by 2050, the International Energy Agency (IEA) estimates that bioenergy production will need to provide more than 20 % of the world's primary energy, or 150 EJ/year (IEA 2008). For comparison, the U.S. Energy Information Administration (EIA) reports global bioenergy production in 2010 of 1.135 EJ/year (electricity 1.127 EJ/year and biofuels .008 EJ/year) (EIA 2013). To support increased bioenergy production to meet the 150 EJ/year production target, biomass feedstock supply systems will need to provide approximately 15 billion metric tons of biomass annually (IEA 2008; Laser et al. 2009).

[1] Based on estimated annual world supply of 146 billion metric tons (Cuff and Young 1980).

L. Nikolaisen
Danish Technological Institute, Taastrup, Denmark
e-mail: Lars.Nikolaisen@teknologisk.dk

T. Ranta
Department of Energy and Environmental Technology,
Lappeenranta University of Technology, Lappeenranta, Finland
e-mail: tapio.ranta@lut.fi

R. Hoefnagels
Energy & Resources, Faculty of Geosciences, Utrecht University,
Utrecht, The Netherlands
e-mail: r.hoefnagels@uu.nl

Currently, world biomass markets and supply chains are not well developed; however, demand is growing, particularly in densely populated areas. The wood pellet market is strong in Europe, which, in 2010, comprised about two-thirds of the total international solid biofuel trade (Lamers et al. 2012). As a result of national policies to increase renewable energy resources, markets are also emerging in China, South Korea, and Japan (Roos and Brackley 2012).

5.1.1 Geographic Incongruity of Resource Production and Energy Demand

A fundamental problem for expansion of bioenergy industries is that regions rich in biomass resources are not necessarily close to the densely populated cities with the greatest energy demands, and the question that often arises is whether bioenergy resources and products can be traded outside their production areas economically and sustainably (Hamelinck et al. 2005) (Fig. 5.1).

5.1.2 Overcoming Geographic Incongruities Within Cost Constraints

As can be inferred from Fig. 5.1, achieving objectives for increased use of renewable bioenergy resources will require that biomass has the characteristics needed to be bought and sold outside of its production areas, or that biomass is "tradable." Tradability is influenced by the reliability of product supply, the existence of a market demand, the opportunity for profitable transactions, the physical transferability of the product, and the guarantee of product quality. These influences are not exclusive of one another, and the more that must be done to a product to improve its tradability, the greater the cost constraint pressures become.

For example, in terms of physical transferability, technologies already exist to preprocess biomass into dense, flowable, storable, and easily transportable feedstocks that can be traded outside of their production areas. Unfortunately, the costs of implementing these technologies often prohibit access to these feedstocks because currently, the costs of improving their tradability keep them from being cost-competitive with the conventional energy products they are intended to replace. As a general guideline based on current states of conversion technologies, it is estimated that delivered biomass feedstock costs need to be near $80 per ton for transportation fuels and $50 to 75 per ton for heat and power. Except for minimally preprocessed feedstocks that are produced in high-yield areas and are cost-effective only for local use, addressing logistics challenges associated with tradability *within cost constraints* is a significant barrier that prevents much of the global biomass resource potential from moving into the market (Fig. 5.2).

Although reliance on local biomass resources is economically viable in a limited number of high-yield scenarios, such as the use of corn stover in the U.S. Midwest

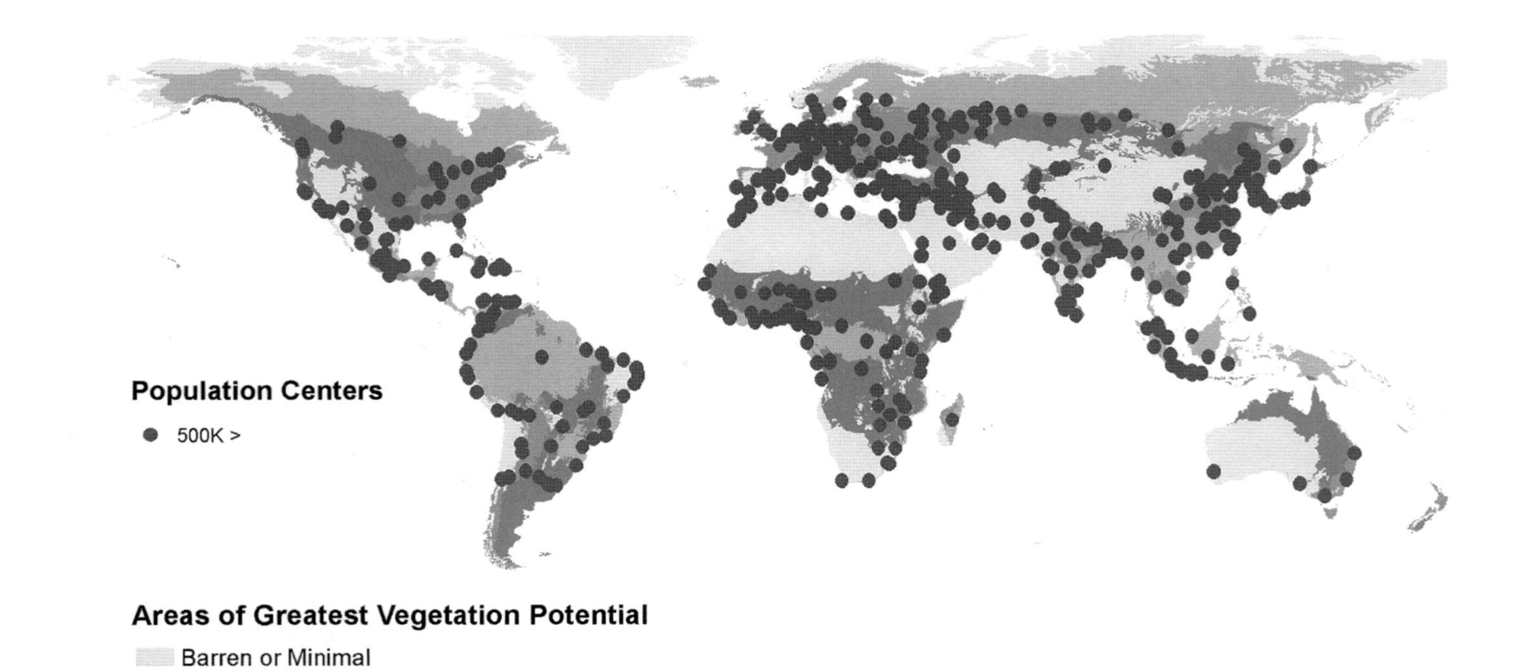

Fig. 5.1 Often the world's regions that are richest in biomass resources (*shaded areas*) are at a distance from the densely populated cities (*black dots*) where demand for energy is greatest, necessitating the transport of bioenergy resources and products outside production areas

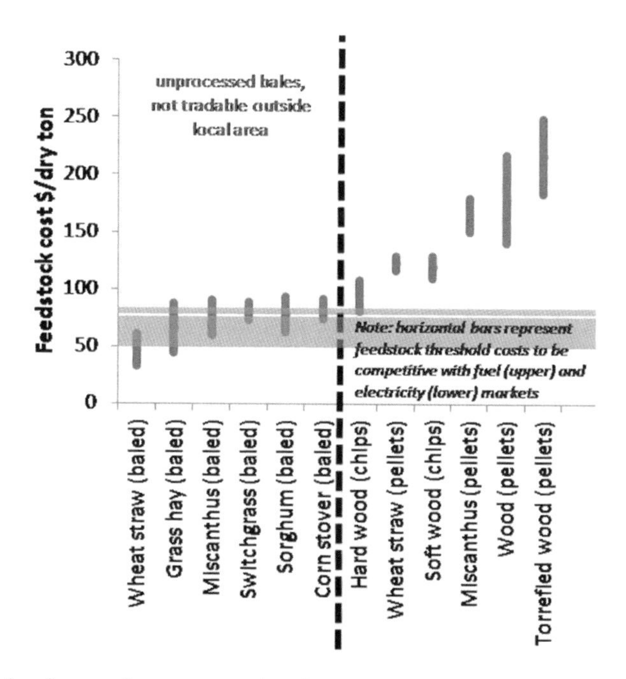

Fig. 5.2 Feedstock costs for unprocessed and preprocessed, tradable bioenergy feedstocks in comparison with feedstock threshold costs that allow biomass to be cost-competitive with fossil-fuel-based counterparts (Pami 2012; Amosson et al. 2011; Saskatchewan Forage Council 2011; Gustafson et al. 2011; Peng et al. 2010; Wood Resources International LLC. 2011; Wright et al. 2012; Nolan et al. 2010; English et al. 2013; Zhang et al. 2010; Sikkema et al. 2010)

to produce renewable transportation fuels, these scenarios do not supply sufficient biomass to support a bioenergy industry capable of achieving the 150 EJ/year production target. To expand global bioenergy production at a scale that substantively captures environmental and energy security benefits, technological advancements are needed that:

1. address the product's physical transferability challenges that currently restrict the volume of resource available
2. ensure specifications meet conversion process performance requirements, and
3. allow bioenergy products to be cost-competitive with their fossil-fuel-based counterparts.

This chapter explores these three themes in terms of feedstock "logistics", "quality", and "market" characteristics, with emphasis on how technological advancements can influence these characteristics by addressing transport and logistics challenges.

From a historical perspective, we consider the development of international grain and flour trading systems, with emphasis on how technology was used to improve logistics, quality, and market characteristics, and, in turn, facilitate significant industry expansion. We discuss similar implementation of technologies to improve tradability of biomass resources, and the tipping point characteristics that both meet logistics and quality requirements for tradability but currently fall short of satisfying international tradability because of economics requirements. Some of the logistics challenges specific to intercontinental transport are presented. Finally, we discuss the potential of using advanced preprocessing technologies in combination with densification and feedstock formulation to increase resource availability, facilitate long-distance trade within cost constraints, and achieve global bioenergy production targets.

5.2 Logistics, Quality, and Market Challenges that Inhibit Global Bioenergy Industry Expansion

The projected future demand for bioenergy presents economic development opportunity for regions that are rich in biomass resources: such regions represent potential "net suppliers" of renewable bioenergy. However, capitalizing on these opportunities by transporting biomass from areas of high production to areas of high energy consumption is inhibited because of logistics challenges influencing delivered feedstock cost.

Currently, feedstock cost is the major determinant of the viability of commercial-scale bioenergy production (Kenney et al. 2013). While many factors contribute to delivered feedstock cost (e.g., grower payment/stumpage fees, transportation, preprocessing biomass in preparation for conversion), long-distance biomass trade incurs additional costs largely due to inherent biomass characteristics and complex logistics (Hamelinck et al. 2005).

5.2.1 The Problem with Biomass

Raw biomass often lacks the qualities that enable it to be traded outside of its local production area, such as uniformity in format and chemical content, high bulk density, flowability within high-volume handling infrastructures, and storability. Lignocellulosic biomass exists in a variety of forms, including herbaceous biomass (such as agricultural residues and energy crops) and woody biomass (such as whole trees and residues). Because biomass is dispersed, it is often processed to facilitate handling prior to transport from the field. This introduces further variability in the biomass by creating a variety of formats, including round bales, square bales, chips, etc. This variability in chemical composition and physical characteristics (such as bale type, particle size and range, etc., generally termed "format") creates challenges to bioenergy producers that often have conversion processes that are highly sensitive to format and composition. In addition, the aerobically unstable biomass can rot during storage, resulting in dry matter losses and further degradation of quality.

Transportation and handling of low-density, cohesive, and degradable biomass materials are substantial barriers to a long-distance biomass feedstock supply system. If biomass could flow through conventional feeding, conveying, and storage systems, logistical costs could be greatly reduced. Unfortunately, many of the existing handling systems are optimized for small granular materials, such as food grains and minerals. These materials are non-cohesive, have small particle sizes and distributions, high densities, and are resistant to compression. In contrast, lignocellulosic feedstocks are often cohesive and have large particle size variations and low densities. They often are highly compressible, causing them to arch over hopper openings and plug mechanical and pneumatic conveying systems (Westover et al. 2011).

These challenges can be addressed by either (1) designing transportation, handling, and storage systems that accommodate the variety of types and formats of raw biomass or (2) formatting the biomass to be compatible with existing infrastructure. Leveraging existing transportation and handling infrastructure to move densified biomass is more feasible near-term and could potentially reduce supply chain costs. High-capacity, efficient, and effective technologies already exist that are capable of moving massive quantities of bulk solids and bulk liquids, such as rail and barge systems used in the grain, coal, and petroleum industries. These transportation modes have some restrictions, such as access to rail lines or waterways, as well as vessel capacity and minimum shipping distances; however, formatting biomass to be compatible with these infrastructures may offer significant cost savings over large distances.

For comparative analysis, it is useful to consider the development of other high-volume global industries— in particular how they responded to geographic imbalances of supply and demand and long-distance logistics challenges. Bioenergy feedstock supply system development is occurring in a very different environment than that of the fossil-fuel resources they are intended to replace or the agricultural commodities they are produced alongside. These industries evolved and became increasingly efficient in parallel with the evolution of transportation and communication networks. The liquid and solid systems developed to manage these commodities are the mature infrastructure that bioenergy trade can leverage for near-term efficiency and cost-effectiveness. For solid bioenergy feedstock trade, the grain trading infrastructure is a promising supply system model (Hess et al. 2009) and insightful case study.

5.2.2 Case Study: Global Grain Trade System

The physical characteristics of conventional energy feedstocks and many agricultural commodity products both enabled and helped shape today's transportation and handling infrastructures (Velkar 2010). Key points in the evolution of the global grain system provide useful illustrations of the influence of physical characteristics on inhibiting and enabling trade. For example, the shift that occurred in the first half of the 1800s in Chicago, Illinois, from transport of labor-intensive, sacked grain lots to agglomerated bulk-flowable lots, enabled the increased flow of grain so dramatically that the Chicago Board of Trade was uniquely positioned to respond to the expanded foreign demand for grain in the 1850s (Cronon 1991). The infrastructure developed during that time to transport and handle "the golden stream" of grain became the model that allowed grain to flow throughout the world.

In parallel to development of the grain industry infrastructure, advancements in communication both facilitated and necessitated standardization, certification, and regulation of grain product characteristics throughout the supply chain to assure that buyers received the appropriate quality of product for their needs and paid a fair price (Cronon 1991). As communication infrastructure developed following the advent of the telegraph, and information about quality and price could be shared more broadly and rapidly, opportunities for profit were more easily identified. The previously isolated economies of producers in different regions became more unified, as did prices offered for grain (Cronon 1991).

In the latter half of the century, the grain industry would undergo another major shift. The same advances that had helped integrate world grain markets also made local markets susceptible to uncertainties of competition, specifically volatility in market demand and price. In response to these uncertainties, the grain milling industry focused on preserving quality, identifying the best markets for different types of grain, and adding value to the grain product. Fractionation, or staged milling, of grain had a notable impact on increasing trade, as experienced in Minneapolis, Minnesota, when flour production rose from 200K barrels in 1870 to seven million barrels 20 years later and 20.4 million barrels by 1915 (Lass 1998; Storck and Teague 1952).

Industry leaders adopted technological and business innovations to overcome the natural disadvantages of the grain's physical and chemical characteristics, such as inconsistent protein content or incomplete blending of gluten and starch, which resulted in rapid rancidification (Storck and Teague 1952; Velkar 2010). Then, milling processes were re-engineered through adoption of technologies that improved both flour quality and shelf life, such as the middlings purifier, which removed husks from the flour, and the gradual-reduction process, which used a series of rollers to gradually pulverize the flour and integrate the gluten and starch. These technologies were scalable and in Minneapolis allowed milling capacity to grow from 490 metric ton/day (4,000 barrels) in 1880 to 1,470 metric ton/day (12,000 barrels) in 1905 (Danbom 2003). The large mills' size and improvements in product quality gave them (1) significant advantages to control cost of production and (2) influence on domestic and foreign markets.

5.2.3 Emulating Grain and Flour Industry Approach to Overcome Natural Disadvantages of Biomass

The same innovations that were implemented to overcome grain's natural disadvantages also increased supply chain costs, but the advantages of economies of scale and the value added to the product allowed grain and grain products to achieve market equilibrium. Fossil-fuel-based feedstock supply chains have addressed similar challenges and enjoy the benefits of a century (or more in the case of coal) of a well-established logistic infrastructure, which allows for low-cost and efficient transport of these energy carriers.

This is not yet the case for international biomass trade. Traders from Latvia, Portugal, Spain, and Sweden remark that the logistical requirements for feedstock collection often severely limit biomass supply, which, in turn, drives up prices. Transportation of preprocessed feedstock product is also problematic, especially if the conversion facilities are not close to waterways or railroads, and truck transportation over large distances is required. The cost of transportation over waterways also creates challenges, depending on the cargo size and port handling efficiencies (Wilmsmeier et al. 2006).

There are technological solutions that can overcome the natural disadvantages of biomass and fulfil logistical requirements. Consider torrefaction, for example. Torrefaction is a mild thermochemical treatment that takes place under atmospheric conditions and at temperatures ranging from 230–300 °C (Deutmeyer et al. 2012). This treatment has the potential to address several logistics challenges, such as improving bulk density and energy density and thereby reducing downstream transportation costs. Torrefaction also provides other benefits, such as lowering biomass moisture content while increasing material stability and increasing grindability (Tumuluru et al. 2011a). Torrefaction may also lower supply-chain cost by increasing energy density, particularly over large distances. It has been shown to be technically feasible for commercial and residential combustion and gasification applications (Tumuluru et al. 2011a). Upgrading feedstocks by improving energy density and oxygen-to-carbon ratio of the feedstock may make the product more valuable to bioenergy producers and increase demand (Deutmeyer et al. 2012). Particularly, using torrefaction to make biomass as similar to coal as possible would facilitate biomass co-firing in coal plants.

There are challenges for this approach that require further research. For example, torrefied biomass can be difficult to densify, requiring the assistance of a binder to receive high durability. Pelleting torrefied biomass also consumes more electricity than does pelletization of nontorrefied wood. Torrefaction is also limited in its ability to address feedstock property concerns for specific conversion methods, such as the high ash content in straw and some wood wastes, which can cause fouling or corrosion in thermochemical conversion reactors. Combining torrefaction and densification with other pretreatment technologies might mitigate these issues; however, the additional costs of preprocessing torrefied, pelleted biomass to address logistics challenges do not allow the feedstock to be cost-competitive with fossil-fuel-based counterparts (Fig. 5.2).

5.2.4 Implications for Bioenergy Feedstock Trade

The implications for bioenergy feedstock trade cannot be overlooked. Today's energy markets are based on tradable, low-cost, energy-dense feedstocks. Highly efficient and complex transportation and communication systems already exist.

The technologies for effectively addressing the tradability challenges of bioenergy resources also exist, but the advantages of scale and value-added products often fall short of achieving market equilibrium for many types of biomass resources (Fig. 5.2). For this discussion, a comparison of the properties that influence a product's tradability is adapted from van Vactor's (2004) dissertation on energy markets and commodity characteristics (Table 5.1).

The effectiveness of efforts to address feedstock logistics challenges, while also meeting consumer's quality requirements within cost thresholds, is a significant indicator of a feedstock's tradability and likelihood of market success. For example, the least preprocessed feedstock product shown in Table 5.1, forest residues, is not tradable outside its production area because of multiple logistics, quality, and market deficits. The most preprocessed feedstock product, torrefied wood pellets, overcomes critical logistics and quality deficits but is not economically feasible; thus, it has severe market characteristic deficits that restrict its tradability.

A fundamental requirement of expanding the bioenergy industry and enabling biomass trade is to balance the costs of transportation and preprocessing with feedstock value so that delivered feedstock costs allow access to more of the globally available biomass resource. A historical review of the global grain trading system demonstrates that this increases product accessibility; for example, following a reduction in cost, the grain industry experienced a rapid expansion (Velkar 2010). The proportion of British population consuming wheat increased throughout the nineteenth-century compared to consumption of other cereals; in 1800, about two-thirds of the population of Great Britain were estimated to have been consuming wheat, whereas by 1900, wheat consumption had become nearly universal, and consumption of oats and barley declined (Velkar 2010). This expansion was a result of many factors, but greatly enabled by reduction in price (Velkar 2010). A reduction in biomass delivered cost, enabled by cost-effectively addressing logistics challenges while meeting conversion facility specifications, could potentially enable the biofuels industry in a similar fashion.

5.2.5 Maritime-Specific Transport and Logistics Challenges

In general, mobility and durability are the principal logistical challenges that negatively impact feedstock delivered costs and restrict industry expansion. As previously discussed, central to these challenges is the poor flowability characteristics of most raw biomass (Westover et al. 2011). Bales, for example, must be individually (or in very limited numbers) loaded onto a truck from the field and driven to the bioenergy producers. The low bulk density of biomass prevents the truck from meeting its weight limit, further contributing to operational inefficiencies (Hess et al. 2009).

These challenges are magnified when maritime shipping becomes part of the supply system activities, as densities of common maritime cargoes vary from 0.6 tons/m^3 for light grains to 3 tons/m^3 for iron ore (Bradley et al. 2009). Additionally, maritime shipping presents unique transport and logistics challenges. Excessive

Table 5.1 Comparison of logistics, quality, and market characteristics of woody feedstocks–having undergone increasing levels of preprocessing to improve their tradability–in relationship to coal and crude oil competitors

		Forest residue	Wood chips	Wood pellets	Torrefied pellets	Coal/ crude oil
Logistics						
Divisibility	Equally and cheaply divisible; transferrable in small and large lots	✓	✓	✓	✓	✓
Non-hazardous	Low risk of danger if mishandled (affordable risk mitigation may be in place)[a]	✓	✓	✓	✓	✓
Mobility	Easily, efficiently, and affordably transported in existing infrastructure[b]			✓	✓	✓
Durability	Resistant to depreciation or degradation during storage and handling				✓	✓
Quality						
Substitutability	Can be consumed by multiple buyers, produced by multiple producers, or substituted for/by multiple products		✓	✓	✓	✓
Homogeneity	Consistent chemical and physical properties		✓	✓	✓	✓
End-use compatibility	Meets user-defined quality specifications at industrial scale			✓	✓	✓
Market						
Adequate demand	High product acceptance/ demand (market or policy driven)[c]			✓		✓
Reliable supply	High-volume production capacity		✓	✓		✓
Information transparency	Broadly communicated pricing, contracting structures, quality standards, and industrial regulation		✓	✓		✓
Market equilibrium	Cost-competitive with fossil-based market share					✓

[a]Each product has some risk associated with transportation and logistics (i.e., self-ignition; toxic off-gassing, dust explosion)

[b]Mobility can involve product attributes and infrastructure

[c]Note that the wood pellet industry is gaining momentum because there is market demand among bioenergy producers, mostly in Europe, who are willing to pay a high price for the feedstock due to subsidies or tax incentives (for example, carbon credits)

shipping costs are considered a major barrier, often surpassing the cost of customs duties (UNCTAD 2011). The cost of shipping biomass depends on many factors, including demand for shipping in general, reliability of biomass resources, shipping capacity, port efficiency, and feedstock characteristics (see Bradley et al. 2009; UNCTAD 2011 for further details).

One logistical challenge that is introduced with maritime transport is compatibility of biomass format with infrastructure at ports, which are the gateways for the international distribution of cargo. Ports have cargo loading/unloading equipment, storage for goods, as well as connection points for various transportation modes. The infrastructure and logistical management at ports are crucial to the efficient material loading and unloading, and therefore shipping costs (Bradley et al. 2009). Generally, the loading and unloading of dry and liquid bulk cargoes are suited for high-tech mechanized and computerized handling, and the specific practice depends on the average volume of trade of a certain type of cargo in the respective port (Bradley et al. 2009).

There are safety challenges associated with maritime transport of biomass. Wood pellets, for example, are safe when in bags; however, when shipped in bulk in large volumes, pellets are classified as hazardous material due to off-gassing of high levels of CO, CO_2 and CH_4, as well as spontaneous combustion potential (Bradley et al. 2009). Char can auto ignite into a smouldering fire when exposed to air or oxygen (Bradley et al. 2009). Mitigation strategies exist; however, the risks must first be well understood in order to implement the appropriate strategy.

5.3 Overcoming Logistics, Quality, and Market Challenges

As demand for bioenergy increases its share of world trade, supply systems that facilitate efficient transport of biomass from the field to the point of conversion must be developed. There are a variety of approaches to accomplish this, such as leveraging or adapting existing infrastructure; implementing technologies to make feedstocks more infrastructure-compatible; and implementing a combination of preprocessing, densification, and formulation technologies to improve feedstock physical, chemical, and economic performance within the supply system and conversion processes.

5.3.1 Leveraging or Adapting Existing Infrastructure

Where transport over rail or water is not possible, more efficient use of trucks, improvement of roads, and optimization of intermediate storage facilities or terminals may be viable options to overcome these logistic barriers. This is already occurring in some areas; for example, in Finland and Sweden several projects are underway to build up large biomass terminals. In the U.S. state of Georgia, an export terminal once used for the paper industry has been retrofitted to accommodate 1.35 million metric ton/year wood pellet export capacity (Geiver 2012). Similarly, the Port of Tyne in England is building new facilities for handling, storage, and transportation of wood pellets (Simet 2013).

5.3.2 Implementing Technologies to Make Feedstocks More Infrastructure-Compatible

One approach for implementing technologies to make feedstocks more infrastructure-compatible and lower feedstock cost is to locate preprocessing technologies as near as practical to the production location (Hess et al. 2009; Searcy and Hess 2010). This facilitates change of the feedstock's physical format to improve density, flowability, stability, consistency, and quality. Making the biomass format compatible with existing high-capacity transportation and handling infrastructure, such as that used in the grain, coal, and petroleum industries, will reduce the need for new infrastructure. Producing biomass with these characteristics at a cost conducive to energy production requires the development of new technologies, or improvements to existing ones. There are some key characteristics of biomass that are targeted to reduce supply chain costs, including moisture management, density, and quality.

There are many examples of technologies under development that have the potential to revolutionize the bioenergy industry by reducing biomass cost. For example, new drying technologies are being developed to manage moisture content (e.g., greater than 25 %), a common biomass characteristic (Shinners et al. 2007). Although drying technologies exist, many have significant capital and/or operating cost.

Efforts to increase energy and volumetric density, by preprocessing raw material into formats that are easily transported and handled, will have a significant impact on total costs. Eventually, increased preprocessing capital and operating cost will dilute this savings. However, the net effects as witnessed in the market are still very positive, and represent the single highest cost-reduction possibility along the whole value chain supplying biomass feedstock to bioenergy producers.

Densification via pelletization has been demonstrated to successfully address some logistical challenges and may offer benefits to biochemical conversion platforms as well (Theerarattananoon et al. 2012; Rijal et al. 2012; Shi et al. 2013; Ray et al. 2013). Densification of comminuted wood particles into pellets results in an increase in energy per cubic meter by a factor of four to five in respect to wood chips, which was necessary to enable international trade of low-value wood for energy purposes. However, this is still well below the energy density of competing energy carriers such as coal, and therefore further increases of density are needed.

As suggested by the earlier discussion of torrefaction, there are technologies under development that, in combination, could potentially revolutionize the bioenergy industry. These technologies address barriers associated with economic biomass supply, including a short operational window, low bulk density, high moisture (and therefore instability), and other quality challenges.

By itself, torrefaction is a potentially low-cost method of increasing energy density by reducing the biomass of approximately 20 % of its volatiles and achieving near-zero moisture content (Tumuluru et al. 2011a). While torrefied wood offers a number of benefits over raw wood, torrefied biomass has low bulk density and is not economical to transport over long distances. Further, the brittleness of torrefied biomass can lead to large proportions of explosive dust, requiring the torrefied product to be classified as a hazardous good and resulting in negative cost impacts. Combining torrefaction and densification (i.e., torrefied compacted biomass [TCB] pellets) can increase the energy density of biomass approximately fivefold (Deutmeyer et al. 2012). Combined torrefaction and densification also produces a biomass feedstock better suited for blending with coal, offering improved milling and handling characteristics, and allowing the two to be blended prior to coal milling, which can potentially increase co-firing ratios (Tumuluru et al. 2011a). By addressing feedstock diversity challenges (Deutmeyer et al. 2012; Tumuluru et al. 2011a) and improving supply chain logistics costs for long-distance maritime transport (Fig. 5.3), TCB pellets have the potential to increase available biomass resources by improving supply/demand economics and expanding the quantity of biomass that is available for energy production.

The absolute cost effects in Fig. 5.3a are calculated on basis of freight when employing a Handy size (15,000–35,000 ton) or Handymax size (35,000–58 ton) dry bulk carriers. These absolute figures do not apply to all situations, as shipping markets are volatile and change of vessel size can have significant cost effects as well. Figure 5.3b illustrates such potential savings by increasing both energy and volumetric density of wood by torrefaction and pelletization in comparison to just pelletization on selected routes using appropriate vessel sizes (i.e., Handymax and Panamax [60,000–80,000 ton]). The cost effect comparing white wood pellets at 625 kg/m^3 and 17GJ/mt with TCB pellets at 700 kg/m^3 and 21 GJ/mt is approximately minus 0,76$/GJ, or nearly 13$/mt. The cost savings illustrated are not limited to shipping and can be achieved along the entire logistical chain.

Logistics costs, despite often being charged per weight, are mostly determined by available transportation volume for higher-stowing cargo, while handling is charged purely on a weight basis. Hence, cost advantages of approximately 37 % can be achieved in rail, barge, and oceangoing-ship transport where volume and not weight is the limiting factor in comparison with wood pellets on a per GJ basis, while for loading and unloading, as well as for trucking, a 23 % advantage is realistic since both are calculated or limited by weight (Deutmeyer et al. 2012).

5.3.3 *Implementing Technologies to Manage Feedstock Variability*

One feedstock trade challenge is the variability of biomass, which includes a variety of materials that have a range of chemical characteristics and material formats. This variability is a not only a challenge for conversion processes, but material handling systems as well.

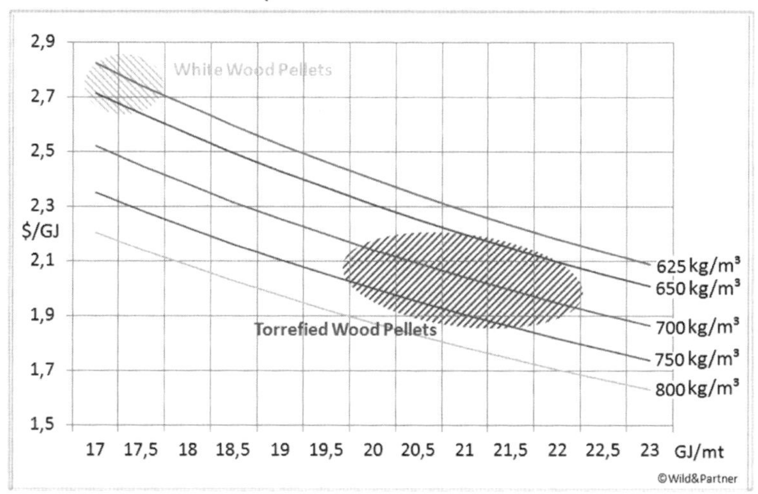

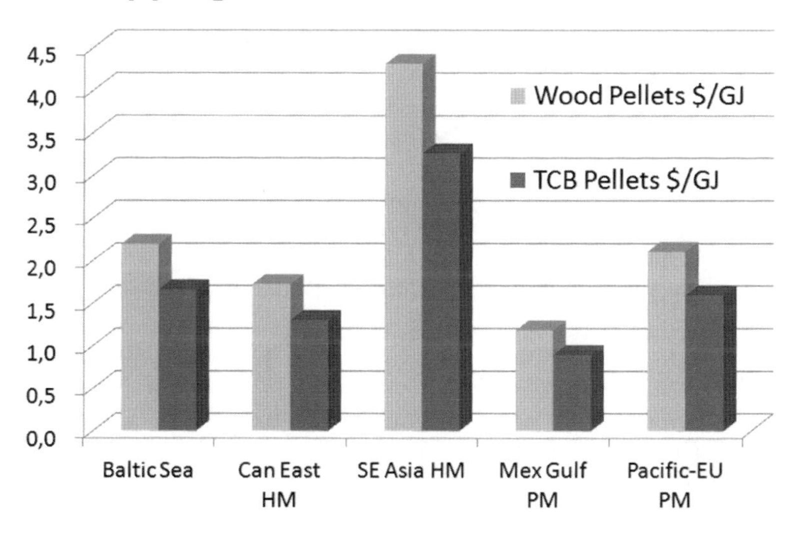

Fig. 5.3 Illustration of reduced shipping cost potential enabled by the use of torrefied compacted biomass (TCB) pellets to increase both energy density and bulk density: (**a**) the cost effect of increased energy density at constant volumetric density of the pellet and (**b**) differences in shipping costs of wood versus TCB pellets (Wild 2012)

In most conventional, locally produced bioenergy production arrangements, the heterogeneity of biomass is the burden of the conversion facility. This can require conversion facilities to have a variety of capital-intensive transportation and handling equipment, which may be economical when relying on a local feedstock resource, but effectiveness over a range of scenarios is limited. Different handling systems, such as conveyors and pneumatic conveyance pipes, have different feed handling requirements. The problem is aggravated by the large variability within each feedstock as well as between the different feedstocks.

Developing equipment systems that will handle the full range of variability of a given region could be extremely challenging, and likely cost prohibitive. By controlling the variability of feedstock parameters, such as particle size (including distribution) and moisture content within and between the feedstocks, the constraints on the handling systems can be greatly relaxed (Kenney et al. 2013). In addition to establishing a limited number of consistent formats, addressing quality characteristics within each format is also an issue. Biomass is highly variable in terms of carbohydrate and ash content, for example, which are two parameters than can have a significant impact on a conversion process. Conversion processes are optimized for certain material specifications, and therefore consistency of in-feed material is also important.

Fractionation of biomass can generally be divided into two types of processes: chemical and mechanical. Chemical fractionation is typically an integral part of the pretreatment process, while mechanical fractionation takes advantage of the variability in biomass characteristics, either whole or ground, to separate the biomass into different streams of varying utility and value. The latter provides an opportunity to improve biomass quality and reduce costs associated with the transport and use of biomass through fractionation. Examples of chemical fractionation include steam explosion, ammonia fiber explosion, and dilute acid hydrolysis. The intent of chemical fractionation is to disrupt structure of the lignocellulosic matrix and/or isolate the cellulose of the biomass to improve the efficiency of enzymatic reactions.

There are many benefits that can be achieved via fractionation. Using mechanical fractionation technology, low-value whole biomass can be upgraded by separating it into multiple streams and co-products with different uses and values. Improvements in size uniformity can enhance flowability, and enable the delivery of on-spec feedstocks to the refineries. Fractions high in ash can be removed to make the feedstock more suitable for thermochemical conversions. The removal of unwanted fractions prior to transport will reduce costs. Mechanical fractionation can also increase the value of the feedstock by concentrating the more desirable elements in the collected fractions. There are several mechanical methods that can be used to fractionate biomass, including fractionation based on particle size, material density, and anatomical or botanical fractionation.

Fractionation technologies have enabled numerous commodity markets, including the international grain industry. The grain milling technology in use around 1870 was relatively unchanged for over a 100 years since steam milling had reduced the industry's dependence on wind and water (Velkar 2010). The use of millstones continued for grinding wheat with limited improvements in the intervening period. This grinding method ensured that the wheat grains were ground thoroughly and as quickly as possible; however, the flour obtained contained a significant proportion of unwanted bran. New developments in milling technology involved improvement and perfection of roller milling techniques, which produced whiter flour with only slightly less yield. The main advantage of this new technology was improved quality, and the whiteness of flour obtained for the same proportion of grains used to produce the coarse 'household' grade flour using the older grinding technology (Velkar 2010). The speed and broad adoption of roller milling was shaped by at least three factors: increasing domestic demand for white flour, unsuitability of softer domestic wheat varieties to the technology, and increase in the imports of foreign flour and hard wheat varieties (Velkar 2010).

There are parallels with the bioenergy industry, which is seeing an increasing demand for biomass to meet long-term biofuels and national renewable energy production goals and, subsequently, the need for quality feedstocks.

5.4 Least-Cost Formulation to Enable Industry Expansion

A feedstock concept designed to increase available biomass resource within cost constraints is least-cost formulation of feedstocks to improve logistics, quality, and market characteristics simultaneously. A formulated feedstock combines various feedstock qualities, types, or constituents to get a blended material that is affordable and meets the in-feed requirements. This approach may combine high-quality, high-cost material that meets or exceeds the in-feed requirements and blend it with low-quality, low-cost material. This process has three objectives:

1. Reduce the overall cost of the material by blending high-cost and low-cost resources
2. Reduce variability of the final blended feedstock by blending to a certain criteria
3. Bring more material into the supply system by enabling use of resources that otherwise would not be suitable for conversion (Muth et al. 2013).

Least-cost formulation enables feedstocks to play an important role in economically and efficiently converting biomass into bioenergy products. It may involve intermediate preconversion steps that break down, clean up, stabilize, and make biomass more reactive to biochemical and thermochemical conversion. This concept combines various preprocessed biomass resources and/or additives to produce an on-spec feedstock that has good physical transferability, durability, and is tradable as a commodity (Muth et al. 2013).

Formulation can also be used to mitigate the effects of undesirable components in raw biomass resources, such as ash. The resulting feedstock will provide consistency and lower costs to bioenergy industries because they can design their processes around a single feedstock that is crafted from numerous, variable resources.

Least-cost feedstock formulation has the potential to overcome logistics and quality challenges for achieving environmental and economic targets, which will require a significant increase in the amount of biomass that is used for bioenergy production. A modeled case study of the effect of least-cost formulation in increasing biomass resource available for bioenergy production is shown in Fig. 5.4 (Muth et al. 2013).

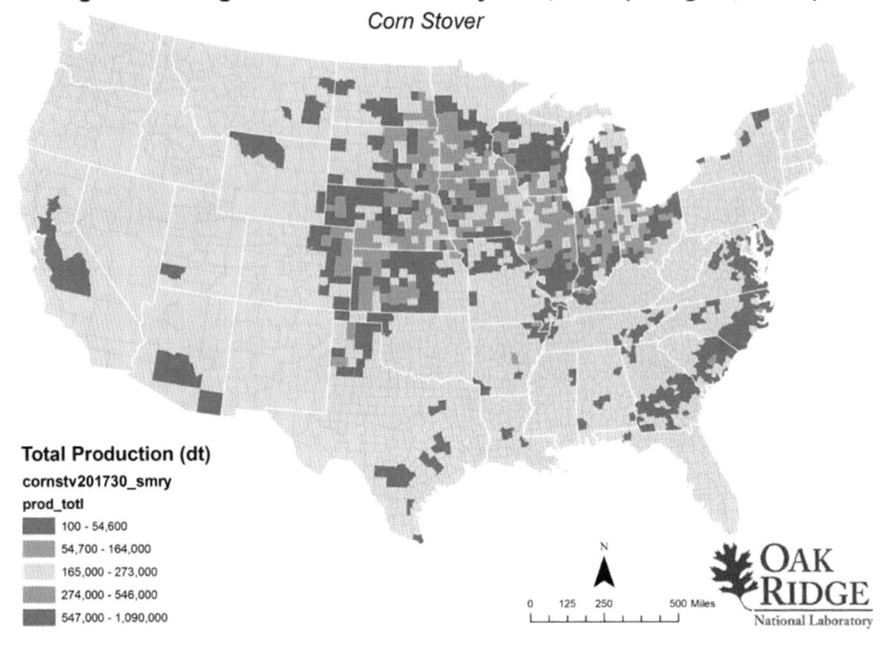

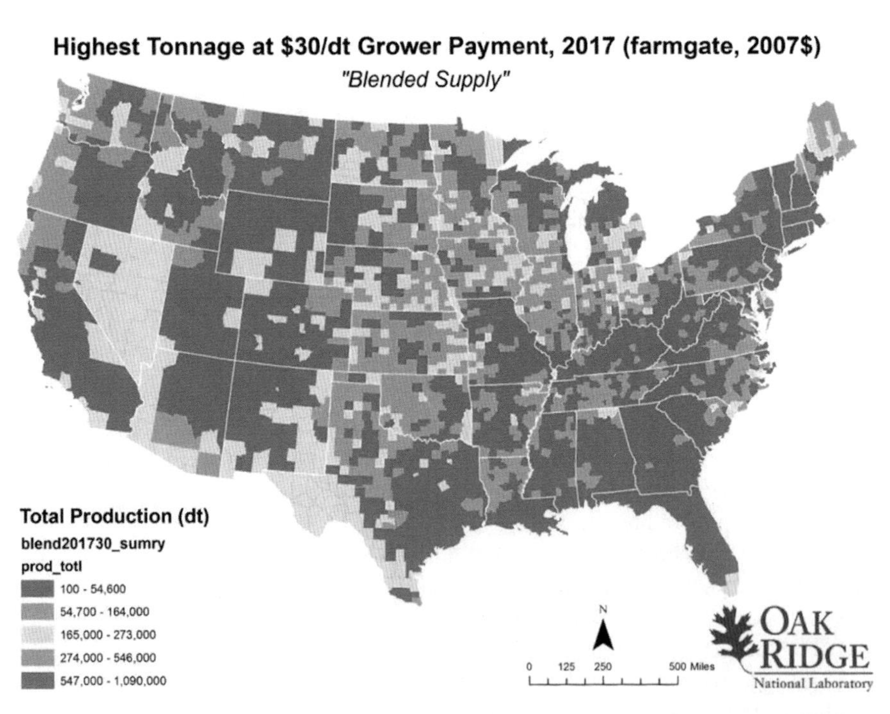

Fig. 5.4 Least-cost formulation increases the biomass resources entering the system by combining high-cost biomass with lower-cost biomass, resulting in a blendstock that meets cost targets. The top map shows feedstock availability in a stover-only scenario (**a**), and the lower map shows the increased availability in a blended supply scenario (**b**), The modeled feedstock blend includes corn stover, perennial grasses, and other biomass resources including thinnings and logging residues (Muth et al. 2013) (Data extracted from the Knowledge Discovery Framework, https://bioenergykdf.net/)

5.5 Conclusions

A barrier to the expansion of this industry is that biomass is not always geographically where it needs to be, nor does it have the tradability characteristics required for efficient handling, storage, and conversion, due to low energy density compared to fossil fuels. Technologies exist that can address these challenges; however, the associated costs often result in a delivered feedstock cost that is not cost-competitive with fossil fuels.

Two key barriers that impact the market success of biomass are mobility and durability. Many biomass forms, such as bales and woodchips, are low in dry matter bulk density. Torrefaction, a thermochemical treatment that drives off moisture and volatiles, produces an energy-dense, stable, and easily transported feedstock product. The product lacks bulk density, which can be improved using densification technologies such as pelletization to increase dry matter bulk density and address feedstock mobility and durability issues that are coupled to long-distance transport and logistics. There are many advantages to densification, including improved handling and conveyance efficiencies, controlled particle size distribution for improved uniformity, improved stability, quality improvements, and improved performance in conversion systems (Tumuluru et al. 2011b). Currently, however, these technologies are often energy intensive and too costly.

As demand for bioenergy increases, the amount of feedstock resources required to support production will be significant. To meet feedstock needs, a typical biorefinery may receive a variety of feedstocks ranging from switchgrass to corn stover to miscanthus to eucalyptus, depending on location and availability. These feedstocks vary widely in composition and recalcitrance, and would require biorefineries to optimize (and possibly re-engineer) their processes for each different type of biomass, thus increasing costs. Complicating this further is that feedstock diversity varies markedly from region to region, and each feedstock within a region varies from year to year based on weather conditions, handling, storage, and crop variety. This will result in different types of biorefineries needed in every region which will further increase costs for construction and operation since there will be no "standard" biorefinery.

Least-cost formulation, in conjunction with mechanical preprocessing and preconversion technologies, offers a promising solution to these issues by combining feedstocks to achieve desired feedstock specifications, reduce undesirable properties, and simplify downstream processing. This approach leverages technology advances to address transport and logistics challenges and convert raw biomass into feedstocks that are easily traded outside their production areas.

References

Amosson, S., Girase, J., Bean, B., Rooney, W., & Becker, J. (2011). *Economic analysis of biomass sorghum for biofuels production in the Texas high planes*. Agrilife Research. Available at: http://amarillo.tamu.edu/files/2011/05/Biomass-Sorghum.pdf. Accessed 30 Jan 2013.

Bauen, A., Berndes, G., Junginger, M., Londo, M., & Vuille F. (2009). Bioenergy – A sustainable and reliable energy source. A review of status and prospects. *IEA Bioenergy: ExCo*:2009:06.

Berndes, G., Hoogwijk, M., & van den Broek, R. (2003). The contribution of biomass in the future global energy supply: A review of 17 studies. *Biomass and Bioenergy, 25*, 1–28.

Bradley, D., Diesenreiter, F., Wild, M., & Tromborg, E. (2009). World biofuel maritime shipping study. *IEA Task 40*. http://www.bioenergytrade.org/downloads/worldbiofuelmaritimeshipping-studyjuly120092df.pdf

Campbell, J. E., Lobell, D. B., Genova, R. C., & Field, C. B. (2008). The global potential of bio-energy on abandoned agriculture lands. *Environmental Science and Technology, 42*, 5791–5794.

Cronon, W. (1991). *Nature's metropolis: Chicago and the great west*. New York: Norton.

Cuff, D. J., & Young, W. J. (1980). *U.S. energy atlas*. New York: Free Press/McMillan Publishing Co.

Danbom, D. B. (2003). Flour power: The significance of flour milling at the falls. *Minnesota History, 58*, 270–285.

Deutmeyer, M., Bradley, D., Hektor, B., Hess, J. R., Nikolaisen, L., Tumuluru, J., & Wild, M. (2012). *Possible effect of torrefaction on biomass trade* (Prepared for the International Energy Agency bioenergy task 40). IEA Bioenergy Task 40. Available at: http://www.bioenergytrade.org/downloads/t40-torrefaction-2012.pdf

English, B. C., Yu, T. H. E., Larson, J. A., Menard, R. J., & Gao, Y. (2013). Economic impacts of using switchgrass as a feedstock for ethanol production: A case study located in east Tennessee. *Economics Research International 2013*, 1–14.

Geiver, L. (2012, December 12). Georgia wood pellet export terminal will be largest in Southeast. *Biomass Magazine*. http://www.biomassmagazine.com/articles/8408/georgia-wood-pellet-export-terminal-will-be-largest-in-southeast

Gustafson, C. R., Maung, T. A., Saxowsky, D., Nowatzki, J., & Milijkovic, T. (2011). *Economics of sourcing cellulosic feedstock for energy production*. Agriculture & Applied Economics Association's 2011AAEA &NAREA Joint Annual Meeting. Pittsburgh, Pennsylvania.

Hamelinck, C. N., Suurs, R. A., & Faaij, A. P. C. (2005). International bioenergy transport costs and energy balance. *Biomass and Bioenergy, 29*, 114–134.

Hess, J. R., Kenney, K. L., Ovard, L., Searcy, E. M., & Wright, C. T. (2009). *Uniform-format solid feedstock supply system: A commodity-scale design to produce an infrastructure-compatible bulk solid from lignocellulosic biomass*, INL/EXT-08-14752. Available at www.inl.gov/bioenergy/uniform-feedstock

IEA. (2008). *International Energy Agency, energy technology perspectives 2008—scenarios and strategies to 2050* (pp. 307–338). Paris: International Energy Agency.

Kenney, K. L., Smith, W. A., Gresham, G. L., & Westover, T. L. (2013). Understanding biomass feedstock variability. *Biofuels, 4*(1), 111–127.

Lamers, P., Junginger, M., Hamelinck, C., & Faaij, A. (2012). Developments in international solid biofuel trade: An analysis of volumes, policies, and market factors. *Renewable Sustainable Energy Reviews, 16*, 3179–3199.

Laser, M., Larson, E., Dale, B., Wang, M., Greene, N., & Lynd, L. R. (2009). Comparative analysis of efficiency, environmental impact, and process economics for mature biomass refining scenarios. *Biofuels Bioproducts & Biorefining, 3*, 247–270.

Lass, W. E. (1998). *Minnesota: A history*. New York: WW Norton.

Muth D. J., Jacobson, J. J., Cafferty, K., & Jeffers, R. (2013) Define feedstock baseline scenario and assumptions for the $80/DT target based on INL design report and feedstock logistics projects. Idaho National Laboratory Joule Milestone ID#:1.6.1.2.DL.4.

Nolan, A., McDonnell, K., Devlin, G. J., Carroll, J. P., & Finnan, J. (2010). Economic analysis of manufacturing costs of pellet production in the Republic of Ireland using non-woody biomass. *The Open Renewable Energy Journal, 3*, 1–11.

Pami. (2012). *Logistics of Agricultural based biomass feedstock for Saskatchewan: For ABC Steering Committee, SaskPower, NRCan*. Research report (Project No. E7810), Humboldt.

Peng, J. H., Bi, H. T., Sokhansanj, S., Lim, J. C., & Melin, S. (2010). An economical and market analysis of Canadian wood pellets. *International Journal of Green energy, 7*, 128–142.

Ray, A. E., Hoover, A. N., Nagle, N., Chen, X., & Gresham, G. L. (2013). Effect of pelleting on the recalcitrance and bioconversion of dilute-acid pretreated corn stover under low- and high-solids conditions. *Biofuels, 4*, 271–284.

Rijal, B., Igathinathane, C., Karki, B., Yu, M., & Pryor, S. W. (2012). Combined effect of pelleting and pretreatment on enzymatic hydrolysis of switchgrass. *Bioresource Technology, 116*, 36–41.

Roos, J. A., & Brackley, A. M. (2012). *The Asian wood pellet markets. General technical report* (PNW-GTR0861, 25 p.). Portland: U.S. Department of Agriculture, Forest Service, Pacific Northwest Research Station.

Saskatchewan Forage Council. (2011). *Saskatchewan forage market report*. Available at: http://www.saskforage.ca/Coy%20Folder/Publications/Forage%20Price%20Report/Forage_Market_Report_Jan_2011-SFC_website_version.pdf. Accessed 30 Jan 2013.

Searcy, E. M., & Hess, J. R. (2010). *Uniform-format feedstock supply system design for lignocellulosic biomass: A commodity-scale design to produce an infrastructure-compatible biocrude from lignocellulosic biomass*. INL/EXT-10-20372. INL. Available at: www.inl.gov/bioenergy/uniform-feedstock

Shi, J., Thompson, V. S., Yancey, N. A., Stavila, V., Simmons, B. A., & Singh, S. (2013). Impact of mixed feedstocks and feedstock densification on ionic liquid pretreatment efficiency. *Biofuels, 4*(1), 63–72.

Shinners, K. J., Binversie, B. N., Muck, R. E., & Weimer, P. J. (2007). Comparison of wet and dry corn stover harvest and storage. *Biomass and Bionergy, 31*, 211–221.

Sikkema, R., Junginger, M., Pickler, W., Hayes, S., & Faaij, A. P. C. (2010). The international logistics of wood pellets for heating and power production in Europe: Costs, energy input and greenhouse gas balances of pellet consumption in Italy, Sweden, and the Netherlands. *Biofuels, Bioproducts and Biorefining, 4*, 132–153.

Simet, A. (2013, January 25). European port to expand wood pellet infrastructure. *Biomass Magazine*. Available at: http://www.biomassmagazine.com/articles/8560/european-port-to-expand-wood-pellet-infrastructure

Smeets, E. M., Faaij, A. P. C., Lewandowski, I. M., & Turkenburg, W. C. (2007). A bottom-up assessment and review of global bio-energy potentials to 2050. *Progress in Energy and Combustion Science, 33*, 56–106.

Storck, J., & Teague, W. D. (1952). *Flour for man's bread*. Minneapolis: University of Minnesota Press.

Theerarattananoon, K., Xu, F., Wilson, J., et al. (2012). Effects of the pelleting conditions on chemical composition and sugar yield of corn stover, big bluestem, wheat straw, and sorghum stalk pellets. *Bioprocess and Biosystems Engineering, 35*(4), 615–623.

Tumuluru, J. S., Sokhansanj, S., Hess, J. R., Wright, C. T., & Boardman, R. D. (2011a). A review on biomass torrefaction process and product properties for energy applications. *Industrial Biotechnology, 7*, 384–401.

Tumuluru, J. S., Wright, C. T., Hess, J. R., & Kenney, K. L. (2011b) A review of biomass densification systems to develop uniform feedstock commodities for bioenergy applications. *Biofuels, Bioproducts, and Biorefining, 5*, 683–707.

United Nations Conference on Trade and Development (UNCTAD) (2011) Review of maritime transport. UNCTAD/RMT/2011. United Nations Publication E.11.II.D.4.

U.S. Energy Information Administration (EIA). (2013). *International Energy Statistics: Renewables*. Available at: http://www.eia.gov/cfapps/ipdbproject/IEDIndex3.cfm

van Vactor, S. A. (2004). *Flipping the switch: The transformation of energy markets*. PhD dissertation, University of Cambridge.

Velkar, A. (2010). *Deep integration of 19th century grain markets: Coordination and standardisation in a global value chain* (Working Papers No. 145/10). London: Department of Economic History, London School of Economics. Houghton Street, London.

Westover, T. L., Searcy, E. M., & Wright, C. T. (2011). *Correlate fundamental bulk rheological properties with mechanical feeding and conveying systems*. Idaho National Laboratory E Milestone ID#:1.3.1.4.D.3.ML.5.

Wild, M. (2012, February 23, 24). *3rd biomass power and trade*, Brussels.

Wilmsmeier, G., Hoffmann, J., & Sanchez, R. J. (2006). The impact of port characteristics on international martitime transport costs. *Research in Transportation Economics, 16*, 117–140.

Wood Resources International LLC. (2011, June). *North American wood fiber review*. Bothell, WA, USA.

Wright, C. T., Kenney, K. L., & Jacobson, J. J. (2012). *Integrated model analysis using field- and PDU-scale data to demonstrate feedstock logistics cost of $35.00 per dry ton for corn stover*. INL TM2012-003-0. INL/MIS-13-28680. INL.

Zhang, Y., Mckechnie, J., Cormier, D., Lyng, R., Mabee, W., Ogino, A., & Maclean, H. L. (2010). Life cycle analysis and costs of producing electricity from coal, natural gas, and wood pellets in Ontario, Canada. *Environmental Science & Technology, 44*, 538–544.

Chapter 6
The Role of Sustainability Requirements in International Bioenergy Markets

Luc Pelkmans, Liesbet Goovaerts, Chun Sheng Goh, Martin Junginger, Jinke van Dam, Inge Stupak, C. Tattersall Smith, Helena Chum, Oskar Englund, Göran Berndes, Annette Cowie, Evelyne Thiffault, Uwe Fritsche, and Daniela Thrän

Abstract As the main driver for bioenergy is to enable society to transform to more sustainable fuel and energy production systems, it is important to safeguard that bioenergy deployment happens within certain sustainability constraints. There is currently a high number of initiatives, including binding regulations and several voluntary sustainability standards for biomass, bioenergy and/or biofuels. Within IEA Bioenergy studies were performed to monitor the actual implementation process of sustainability regulations and certification, evaluate how stakeholders are affected and envisage the anticipated impact on worldwide markets and trade. On the basis of these studies, recommendations were made on how sustainability requirements could actually support further bioenergy deployment. Markets would gain from more harmonization and cross-compliance. A common language is needed as 'sustainability' of biomass involves different policy arenas and legal

L. Pelkmans (✉)
Project Manager Biomass & Bioenergy, Unit Separation and Conversion Processes, VITO NV, Boeretang 200, BE-2400 Mol, Belgium
e-mail: luc.pelkmans@vito.be

L. Goovaerts
Unit Separation and Conversion Processes, VITO NV, Boeretang 200, BE-2400 Mol, Belgium

C.S. Goh • M. Junginger
Faculty of Geosciences, Copernicus Institute of Sustainable Development, Utrecht University, Utrecht, The Netherlands

J. van Dam
SQ Consult, Utrecht, The Netherlands

I. Stupak
Department of Geosciences and Natural Resource Management (IGN), Faculty of Science, University of Copenhagen, Copenhagen, Denmark

C.T. Smith
Faculty of Forestry, University of Toronto, Toronto, ON, Canada

M. Junginger et al. (eds.), *International Bioenergy Trade: History, status & outlook on securing sustainable bioenergy supply, demand and markets*, Lecture Notes in Energy 17, DOI 10.1007/978-94-007-6982-3_6, © Springer Science+Business Media Dordrecht 2014

settings. Policy pathways should be clear and predictable, and future revisions of sustainability requirements should be open and transparent. Sustainability assurance systems (both through binding regulations and voluntary certification) should take into account how markets work, in relation to different biomass applications (avoiding discrimination among end-uses and users). It should also take into account the way investment decisions are taken, administrative requirements for smallholders, and the position of developing countries.

6.1 Introduction

Biomass (solid, liquid and gaseous) is considered to play a key role in future energy supply (Chum et al. 2012). It can contribute to the reduction of greenhouse gas emissions, increasing the energy supply diversity and security, and provide opportunities for local communities, overall a more sustainable fuel and energy supply, in environmental, economic and social terms.

However, to meet its promises, we need to ensure that biofuel and bioenergy deployment happens in a way that respects the three pillars of sustainability, i.e. the reconciliation of environmental, social equity and economic demands, both for domestic and imported biomass. The spectacular growth of biofuel production between 2005 and 2008, driven by country mandates, targets and incentive systems, has triggered a discussion on potential sustainability risks of biofuels. On the one hand, biofuels provide new opportunities for agricultural markets and rural communities; on the other hand, there are environmental, social and economic concerns about the production of biomass feedstocks for biofuels. The discussions on sustainability of biofuels, food versus fuel, and land use change often overshadow potential positive effects such as greenhouse gas (GHG) reduction and economic advantages for communities and countries. The discussion of using solid biomass for bioenergy

H. Chum
Thermochemical Process R&D and Biorefinery Analysis, National Renewable Energy Laboratory, Golden, CO, USA

O. Englund • G. Berndes
Physical Resource Theory, Chalmers University of Technology, Gothenburg, Sweden

A. Cowie
Rural Climate Solutions, University of New England, Armidale, NSW, Australia

E. Thiffault
Canadian Forest Service – Laurentian Forestry Centre, Natural Resources Canada, Succ.Sainte-Foy, QC, Canada

U. Fritsche
International Institute for Sustainability Analysis and Strategy (IINAS), Darmstadt, Germany

D. Thrän
Department of Bioenergy Systems, German Biomass Research Center (DBFZ), Leipzig, Germany

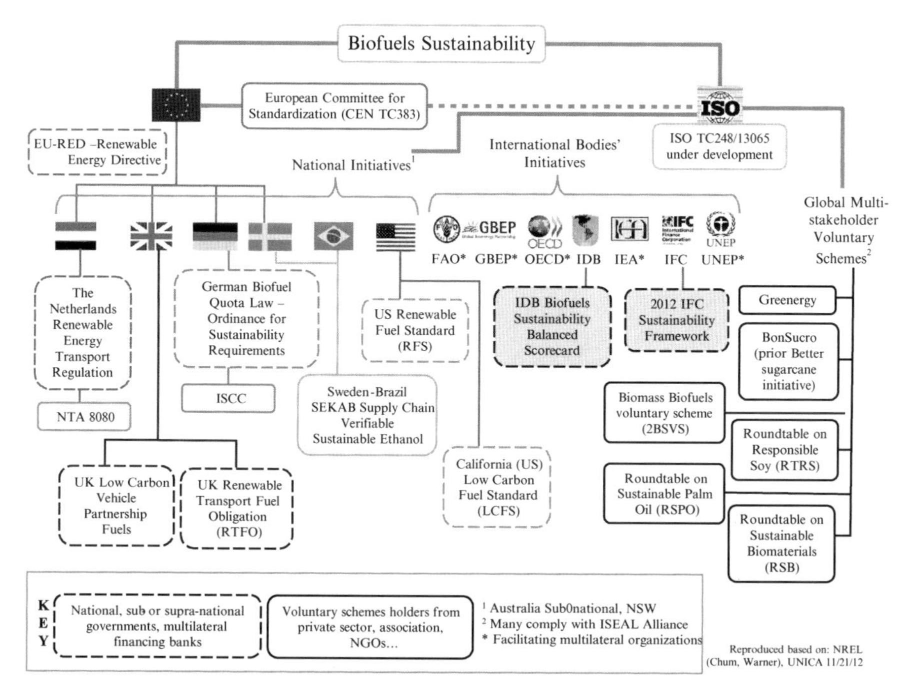

Fig. 6.1 Illustration of some government-led initiatives (in *dashed boxes*) and of sustainability standards in relation to liquid biofuels that were developed over time by a variety of entities (*full boxes*) (Many are organized through voluntary schemes by multiple stakeholders. Others, not displayed, exist for forestry and agriculture, specifically. Scorecards are also used to provide check lists of project submissions to financing by multilateral organizations)

(mainly for electricity and heat generation) follows with some delay the discussions around biofuels for transport. While the debate on biofuels focused on issues of food versus fuel and land use change, the risks for biodiversity and carbon stock loss in forests are prevalent concerns in the debate on solid biomass.

The sustainability of biomass/bioenergy/biofuels can be governed at multiple levels through:

- subnational, national or regional legislation and regulations,
- international conventions and processes,
- jurisdictional guidelines (mandatory or voluntary),
- certification schemes, and
- business systems – Corporate Social Responsibility & Environmental Impact Assessment.

In general one can distinguish between obligatory (regulated) and voluntary systems, which can complement each other. Overall, a large number of initiatives – mainly in the form of regulations and certification schemes – have been developed to ensure the sustainability of land management and biomass production systems, also in relation to markets of biofuels and bioenergy carriers. Figure 6.1 shows an example overview of initiatives which have developed to ensure sustainability of biofuels.

Without going into the details associated with this figure, in general, we see governance initiatives placed on three levels:

- National/regional initiatives developed by regulators; the main examples are regulations in the EU (Renewable Energy Directive), in separate EU Member States (like the Netherlands, UK, Germany, Sweden), and the United States (Renewable Fuel Standard 2) and Brazil;
- Initiatives developed by international bodies such as FAO, GBEP, UNEP, IFC, IDB[1]; and
- Multi-stakeholder voluntary schemes, typically developed by companies and NGOs.

In addition to dedicated biofuel and bioenergy governance systems, various voluntary certification schemes have been in existence since the early 1990s, which aim at ensuring sustainable forest or agricultural management or fair trade. These systems complement most initiatives that were developed for sustainable bioenergy.

6.2 Sustainability Requirements in Legislation

The interest in biofuels and bioenergy production and investment has been largely driven by policies of national governments, both in developed and developing countries, designed to reduce greenhouse gas (GHG) emissions and to reduce dependency on fossil fuel imports. Bioenergy has since long been a significant part of the energy mix in some countries, and was commonly considered an attractive opportunity also for meeting rural development objectives and for job creation associated with the growth of a new industry. However, the recent years' rapid increase in the use of conventional food crops for biofuels, and the proposed linkages to socioeconomic and environmental impacts, spurred an intensified debate about the sustainability of bioenergy. This debate triggered a range of initiatives to develop sustainability standards and certification schemes to account for and monitor sustainability issues intended to reduce the negative unintended consequences of bioenergy expansion.

A number of countries have already been actively engaged in the development of sustainability standards and certification schemes for biofuels and bioenergy, including Australia, Belgium, Canada, Germany, Japan, Korea, the Netherlands, New Zealand, Sweden, Switzerland, the United Kingdom, the United States, and a number of developing countries such as Argentina, Brazil, China, India, the Philippines, South Africa and Thailand. An overview of such initiatives can be found in several publications like Scarlat and Dallemand (2011); van Dam et al. (2010) or O'Connell et al. (2009), but it should be emphasized that since many initiatives have been developed very recently, it is difficult to give a comprehensive overview of them. Countries have adopted policies that encourage the production

[1] FAO = the Food and Agriculture Organization of the United Nations; GBEP = Global Bioenergy Partnership; UNEP = United Nations Environment Programme; IFC = International Finance Corporation (World Bank Group); IDB = Inter-American Development Bank.

and use of bioenergy, mostly related to biofuels, and have set sustainability requirements for production, processing and trade of biofuels, bio-liquids and/or solid and gaseous biomass which must be fulfilled in order to meet present national targets and/or to be eligible for financial support.

The policies that have the greatest impact on large international bioenergy markets are those developed in the European Union and the United States (Pelkmans et al. 2012). A brief overview is presented in the following section.

6.2.1 European Union

The main legislative driving force for sustainability of biofuels and bioenergy in the EU is the Renewable Energy Directive (Directive 2009/28/EC, hereafter called 'RED'). The aim of this legislative act is to achieve by 2020 a 20 % share of energy from renewable sources in the EU's final consumption of energy and a 10 % share of energy from renewable sources in each Member State's transport energy consumption (EC 2009). The RED has set specific minimum sustainability standards for ***biofuels*** (for transport) ***and bioliquids*** (for electricity and heat production) and requirements for their verification that should be met in order to receive government support or count towards the mandatory national renewable energy targets. The sustainability criteria are:

- greenhouse gas (GHG) savings of at least 35 % compared to fossil fuel (to be increased up to 50 % from 2017 and 60 % for new installations from 2018),
- no raw material from land with high biodiversity value, such as primary forest, nature protection areas, highly biodiverse grasslands *(unless it can be shown that biomass extraction is part of a management regime compatible with – or a requirement for – high biodiversity)*,
- no raw material obtained from converted[2] high carbon stock land (continuously forested areas, wetlands or peatlands),
- raw material coming from European agriculture needs to be produced following 'good agricultural practices' as described in the Common Agricultural Policy (CAP).

The compliance to these biofuel sustainability requirements needs to be checked by Member States or through voluntary schemes which have been approved by the European Commission (EC).[3] The EU Member States must also report to the EC on biannual basis on the impact of biofuels and bioliquids on biodiversity, water resources, water and soil quality, GHG emission reduction and changes in commodity

[2] Converted according to the RED = land that had the status of continuously forested areas, wetlands or peatlands in January 2008 and no longer has that status.

[3] Since 19 July 2011, the EC has recognised voluntary schemes for biofuels, applying directly in the 27 EU Member States: ISCC, Bonsucro, RTRS, RSB, 2BSvs, RBSA, Greenergy, Ensus, Red Tractor, SQC, Red Cert, NTA8080, RSPO. http://ec.europa.eu/energy/renewables/biofuels/sustainability_schemes_en.htm

prices and land use associated with biomass production. The RED in itself did not include any definite set of definitions, criteria and indicators related to terms such as "primary forest" and "highly biodiverse grasslands" requiring that these be further examined and defined as part of a comitology process at EU level.

On 17 October 2012, the EC published a proposal to limit global land conversion for biofuel production, and raise the climate benefits of biofuels used in the EU.[4] The proposal contains four major changes:

- Incorporation of biofuels produced from food crops (cereals, sugar and vegetable oil) would be limited to 5 % in terms of energy content out of the target of 10 % of renewable energy in transport by 2020,
- New biofuel plants (post 1st July 2014) should deliver minimum greenhouse gas savings at 60 % compared to fossil fuels emissions,
- Additional support is introduced for "advanced" biofuels produced from non-food feedstocks, such as waste, straw and non-food crops, by weighting more favourably their contribution towards the 10 % renewable energy target,
- The estimated GHG emissions associated with indirect land use changes (iLUC) needs to be reported by Member States and fuel suppliers based on using fixed factors.[5] The high iLUC value for oil crop biofuels puts a high constraint on the role of biofuels from oil crops after 2020.

The EC also expresses the view that in the period after 2020 biofuels produced from food and feed crops, which do not lead to substantial greenhouse gas savings (when iLUC emissions are included), should not be subsidised.

So far the RED sustainability requirements do not apply for ***solid or gaseous biomass*** used for electricity or heat production. However, feedstocks used for the production of solid and gaseous bioenergy carriers (notably lignocellulosic biomass) are expected to also be used for the production of '2nd generation biofuels', which will have to comply with the requirements set for biofuels and bioliquids. It is therefore expected that common requirements or some form of harmonization will be needed.

In February 2010, the EC published a Communication[6] stating that for the moment, there would be no binding criteria at the European level. However, the EC provided a number of recommendations for Member States in order to ensure greater consistency and to avoid unwarranted discrimination in the use of raw materials. Basically, it recommended the use of a similar methodology as that for biofuels for installations larger than 1 MW, with the same sustainability requirements on biodiversity and

[4] COM(2012)595, Proposal for a Directive of the European Parliament and of the Council amending Directive 98/70/EC relating to the quality of petrol and diesel fuels and amending Directive 2009/28/EC on the promotion of the use of energy from renewable sources. October 2012.

[5] Current iLUC emission factors are 12 g CO2eq/MJ for cereals, 13 g CO2eq/MJ for sugars and 55 g CO2eq/MJ for oil crops (for reference, the fossil fuel comparator is 83.8 g CO2eq/MJ). Biofuels made from feedstocks that do not lead to additional demand for land, such as those from waste feedstocks, should be assigned a zero emissions factor.

[6] COM(2010)11, Report from the Commission to the Council and the European Parliament on sustainability requirements for the use of solid and gaseous biomass sources in electricity, heating and cooling. February 2010.

high carbon stock land and a common GHG calculation (with adapted reference as the end use needs to be included as well). The EC is in the process of assessing the implementation of its recommendations to Member States, and the opportunity to have binding EU-wide criteria for solid and gaseous biomass. EC recommendations are expected to be released in 2013.

6.2.1.1 Selected Examples

Germany In 2006, the German Ministry launched a project aimed at defining the basis for sustainability requirements for biofuels. The result was the proposed Biomass Sustainability Regulation (BSR). The draft BSR was released in late 2007, but with the RED in development at EU level, the initiative was abolished. Nevertheless, in the early stages Germany decided to follow the RED requirements and it was the first country to implement the sustainability requirements of the RED in their own legislation. Germany also supported the development of a scheme called ISCC (International Sustainability and Carbon Certification). This system was the first to be recognized at the national level to fulfil the RED requirements (in 2010). A second system, the REDcert, was also recognized later in Germany (Lieback and Kapsa 2011). In 2015 Germany will change from volume quota to CO_2-quota for biofuels. This will put higher emphasis on the GHG balance of biofuels (to be certified), with important economic impact.

Belgium Belgian authorities (at regional level) introduced sustainability criteria into their supporting scheme for renewable electricity in 2006. In the Flemish region, certain biomass streams (e.g. wood (waste) that is still suitable for recycling in board or pulp and paper industry) are not entitled to receive green power certificates as a feedstock for the production of renewable electricity. Also, the energy used for transporting and pre-treatment of the biomass, is deducted from the green power certificates. In the Brussels and Walloon regions, a greenhouse gas balance and reduction compared to the best available natural gas system is calculated to determine the amount of green certificates. All calculations must be validated through an audit by an independent organisation.

United Kingdom Since April 2008, under the UK RTFO (Renewable Transport Fuel Obligation), the Renewable Fuels Agency (RFA) requests fuel suppliers to report on the specific type and origin of biofuels, the compliance of biofuel crops with existing environmental and social sustainability criteria, and the greenhouse gas emission reductions achieved by using biofuels. While there are no strict consequences of not meeting the sustainability criteria, public disclosure may be an important driver for the reporting commercial companies. A similar procedure was implemented for renewable electricity in 2011. From 2011, a well-founded report on the RED sustainability criteria is required for installations larger than 50kWe; from 2013, generators of 1MWe and above will need to actually satisfy the sustainability criteria. This staged approach will also be considered by the Renewable Heat Incentive (RHI).

The Netherlands The Netherlands examined sustainability criteria for all forms and applications of biomass. In 2007, the Cramer Commission published a list of sustainability principles for the use of biomass for energy (fuels, liquids, solid and gaseous). These principles are partially covered in the RED sustainability criteria. The Netherlands are building further on their experience with the Corbey Commission. Based on the *Cramer* principles, the Dutch normalisation institute NEN, developed standards NTA 8080 and 8081 for sustainable biomass for energy purposes (NTA 8080 2009). This is a voluntary system and already used by commercial actors to demonstrate the sustainability of their biomass. The NTA 8080/81 was recently approved by the European Commission as a voluntary system for biofuels and bioliquids. In October 2012, large Dutch biomass users have signed a *Green Deal*. The participating companies will report annually to the government the amounts of biomass they use and how sustainability is demonstrated via certification or verification systems.

6.2.2 United States

US Renewable Fuel Standard (RFS2) The RFS2 defines the volume of different biofuels that have to be blended with conventional fuel between 2006 and 2022 according to the US Energy Independence and Security Act of 2007. The total volume of biofuels mandated in the Renewable Fuels Standard will increase to 36 billion gallons (136 million m^3) in 2022. Each year, obligated parties such as refiners and importers of gasoline and diesel and blenders are required to meet volumetric targets for four broad categories of biofuels: (1) conventional renewable fuels; (2) bio-based diesels, (3) advanced biofuels, and (4) cellulosic biofuels. These biofuel categories are defined based on the nature of feedstock/technology used in production and minimum GHG reduction thresholds obtained. These requirements favour the development of highly efficient biofuel technologies, including 2nd generation biofuels. The definition of 'renewable biomass' in the RFS2 limits the types of biomass as well as the type of land from which biomass may be harvested to produce compliant renewable fuels. The law sets a limit of 15 billion gallon (57 million m^3) for conventional renewable fuel.

All renewable fuel producers must report and maintain records concerning the type and amount of feedstocks used for each batch of renewable fuel produced. Additionally, the producer must report to EPA on a quarterly basis concerning the source of the feedstocks. Renewable fuel producers are required to obtain from their feedstock supplier, and maintain in their records, documents which certify that the feedstock meets the definition of renewable biomass and renewable fuel, describe the feedstock, and identify the process that was used to generate the feedstock.

To track achievement towards the mandate for renewable fuel, EPA established a system of tradable Renewable Identification Numbers (RINs). Upon blending with gasoline, the RIN is detached from producer and used by the blender as proof of traded renewable fuel or sold to another obligated party. EPA also sets the required volumes of biofuels each year.

The California Low Carbon Fuel Standard (LCFS) State-level legislation in the US, such as the California's Low Carbon Fuel Standard, is also largely based upon reporting requirements using default carbon intensity values established per type of biofuel, although other technologies such as electric vehicles can be used. The California Low Carbon Fuel Standard (LCFS) is a standard that aims to reduce GHG emissions from the transportation sector in California by at least 10 % by 2020, using a technology-independent life cycle approach. These emissions include not only tailpipe emissions but also all other associated emissions from production, distribution and use of transport fuels. The calculations include indirect land use change (iLUC). The California Air Resources Board (CARB) calculated current carbon intensities of various fuel pathways and sub-pathways and listed them in lookup tables. Each additional facility and pathway approved is then found in the registered facility information, which is added to other already registered fuels. The LCFS convened a working group relative to the iLUC factor and this factor will be modified in legislation in the future.

6.2.3 Latin America

In order to address potential negative environmental and social impacts of bioenergy production, several sustainability initiatives have been established in Latin America during recent years. Such efforts have been initiated by stakeholders from the industry, as well as by Latin American governmental bodies. Most sustainability initiatives addressing feedstock production for food, feed and biofuels operate on a voluntary basis. Some are embedded in legislation, particularly in Brazil. Some examples:

- Brazilian agro-ecological zoning for sugarcane – On a national level in Brazil, there is an agro-ecological zoning for sugarcane, including specific requirements regarding appropriate soil and climate, with no or low irrigation requirements, and low slopes for mechanized harvesting and reduced atmospheric emissions. Investors who do not respect this zoning are not eligible for getting loans from public institutions. A similar system is currently being developed for palm oil.
- The Sao Paulo State Green Ethanol programme – An applied tool of the Green Ethanol programme is the agro-environmental sugarcane zoning in the State of São Paulo. This tool is a map with several layers identifying potential sugarcane expansion areas and protected areas.
- The Social Biodiesel Programme in Brazil – The objective is to redistribute wealth, fight against rural poverty and to improve living conditions for poor farmers in north-eastern Brazil. Biodiesel companies that use and buy feedstock at fair prices from smallholders and family farmers gain tax benefits from the state. The programme did not meet its ambitious targets to promote family farmers and alternative feedstock so, as a result, the Brazilian biodiesel market is currently dominated by large-scale soy production.

Sustainability requirements for biofuels and bioenergy in legislation have been steadily implemented in the past years. These are now starting to have the anticipated effects in the field and on international markets.

In general, these (supra)national regulations address the environmental and ecological issues related to biofuel production, such as (1) the climate change mitigation potential of biofuels by requiring a certain percentage in reduction of lifecycle GHG emissions compared to a fossil-based fuel, and (2) preservation of existing organic carbon stores and biodiversity by stating that biofuel production should not cause conversion of land with high carbon stock or high biodiversity value.

Social issues are covered in a different (and somewhat more limited) way, e.g. by setting reporting requirements on social sustainability addressing food availability and price, as well as workers' rights and land access and ownership rights.

The advantage of these national/regional standards is that they are well tailored to local/regional issues. However, initiatives are not always comparable with regards to the overall structure, definitions used, specific sustainability requirements, reporting methodology and reporting requirements; for example, there are differences in the type of biomass/biofuel/bioliquids included, time frame, GHG emission reduction requirements, the GHG emission reduction calculation methodology and the way iLUC is incorporated. As a result, this situation can be confusing to actors in the marketplace and lead to barriers for international trade.

6.3 Voluntary Certification Systems

6.3.1 Introduction

Sustainability certification exists for a wide range of products, addressing good resource management and responsible entrepreneurship. These are generally performance-based schemes aiming to achieve a certain standard, and include a number of principles, criteria and indicators designed to verify compliance. Certification systems have become available for almost all feedstock and products covering parts of, or the complete, supply chain – from production and processing to trade of biomass and biofuels. Some of these systems exist on a national level, and others are internationally recognized and applicable. Certification schemes enable actors along the supply chain and involved with trade to attest that land management and biomass production and procurement practices comply with regulations and requirements regarding sustainable biomass or bioenergy. Due to the fact that these systems have been developed with different interests and priorities (e.g. by governments, NGOs, companies), the scope, approach and complexity vary from scheme to scheme. Certification systems have a number of similarities in terms of coverage of sustainability issues/principles, but there is a variation in the way compliance with standards is measured, i.e. different sustainability criteria and indicator systems and monitoring procedures exist.

A variety of schemes has become operational for the production, processing and trade of biomass, with the most prominent ones relevant for bioenergy markets being:

- **Forest certification systems:** The first implemented forest certification scheme was the Forest Stewardship Council (FSC). The FSC sets international principles for sustainable forest management, and local stakeholders develop region-specific standards. Other schemes followed, with PEFC as one of the larger recognised international certification organizations, endorsing national-level schemes based in more than 30 countries. In general, each of these PEFC schemes differs in how sustainable forest management is defined, but our review indicates they seem to have somewhat similar chain-of-custody standards, although some differences can be found. The PEFC has not mandated one set of international principles but does have a mechanism for evaluating if schemes seeking PEFC endorsement are in compliance with a 'harmonized' set of standards (Stupak et al. 2011). While FSC and PEFC schemes are used to certify the sustainable management of forests from which bioenergy feedstocks are harvested, neither were originally developed for biofuels/bioenergy applications. These schemes also do not include binding limits for GHG emissions, nor do they include the complete production chain or quality of air issues. They do address water and soil quality/conservation, and include biodiversity and workers and land rights.

- **Agricultural certification systems:** Most of these systems are designed for the certification of organic products to be used for a wide range of end-uses (food, feed, energy), like SAN/RA and GlobalGAP. Some focus on a specific crop, like RTRS (soy), RSPO (palm oil) and Bonsucro (sugar cane). As for forestry certification, these agricultural schemes include environmental, economic and social aspects; soil conservation is addressed in all schemes; and air quality is only covered in RSPO and social aspects (workers' rights and land rights) are not included in GlobalGAP. The crop specific schemes, RTRS, RSPO and Bonsucro, have recently been extended to also include specific biofuels or bioenergy related issues, i.e. GHG emissions and carbon conservation, so that they are recognized as voluntary scheme for biofuels by the European Commission.
- **General biofuel/bioliquids certification systems:** A number of dedicated certification schemes for biofuels/bioliquids exist (e.g. ISCC, RSB, REDCert, 2BSvs). Most of them have been developed to show compliance with the European RED requirements. These are more generic standards which cover a wide range of feedstocks to be used for biofuels or bioliquids. They cover the same aspects as the crop dedicated agricultural schemes, although the approach differs; for example, these schemes require a specific GHG reduction target compared to fossil fuel instead of general GHG improvement requirements. On the other hand they generally exclude requirements on e.g. fertilizer applications, tillage, labour conditions and so on.
- **Wood pellet certification systems:** The first private standards for wood pellets for energy production included the Green Gold Label (GGL) and the Laborelec system, which were developed to comply with (anticipated) national legislation and customers demand. These are mainly Chain-of-Custody (CoC) standards for product verification. They allow the use of other schemes to comply with the sustainability criteria set out in the standard (e.g. FSC, PEFC, including e.g. CSA, SFI). Currently a consortium of large pellets buyers have formed an initiative called 'International Wood Pellet Buyers' (IWPB) to streamline their quality and sustainability requirements to facilitate trade within the sector (IWPB 2012).

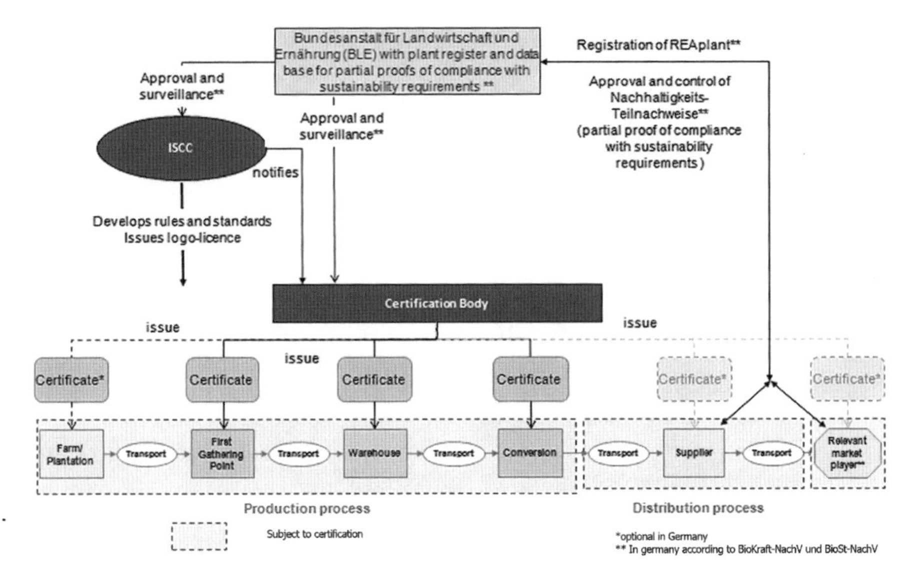

Fig. 6.2 Overview on processes and responsibilities in the ISCC scheme (Note that the involvement of a government body BLE is untypical and generally not the case for other voluntary systems) (ISCC 2012)

Certification enhances the relationship between different stakeholders as a result of verification and certification requirements. It requires stakeholders to communicate with each other on different levels, both during development and improvement of the certification schemes and the implementation (CoC requirements and audits), and thus improves awareness. Figure 6.2 indicates relationships and CoC processes in the ISCC scheme.

Certification affects various market actors differently. The supply side is pushed towards certification to improve trade and gain credibility from the demand side, i.e. buyers and other organisations like NGOs. Both groups of stakeholders thus catalyse the development and implementation of certification schemes.

6.3.2 Implementation of Relevant Schemes

The IEA Bioenergy inter-task study ‘Monitoring sustainability certification of bioenergy’[7] looked at the implementation process of sustainability certification of bioenergy. A list of relevant and representative schemes was selected, also on the basis of relevant trade flows for biofuels/bioenergy. The most important feedstocks in terms of trade flows for energy are ethanol from sugarcane (mostly from Brazil), biodiesel

[7] Results available at http://www.bioenergytrade.org/publications.html. See also Goovaerts et al. 2013, Stupak et al. 2013, Goh et al. 2013 and Pelkmans et al. 2013.

Table 6.1 Selected certification schemes for analysis in the IEA Bioenergy inter-task study (Goovaerts et al. 2013)

Sector	Schemes
Forestry	FSC (Forest Stewardship Council)
	PEFC endorsed schemes, such as SFI, CSA-SFM, ATFS[a] in North-America, PEFC Finland, Sweden, Germany or France in the EU, CertFor (Chile), CerFlor (Brazil), FCR (Russia)
Agricultural crops	GlobalGAP (worldwide standard for Good Agricultural Practice)
	SAN/RA (Sustainable Agriculture Network/Rainforest Alliance)
	CSBP (Council on Sustainable Biomass Production) in the United States
	Bonsucro (sugarcane)
	RSPO (Roundtable on Sustainable Palm Oil)
	RTRS (Round Table on Responsible Soy)
Biofuels (general)	ISCC International Sustainability and Carbon Certification)
	RSB (Roundtable on Sustainable Biofuels)
	2BSvs (Biomass Biofuel Sustainability voluntary scheme)
Wood pellets (for energy)	GGL (Green Gold Label, developed by RWE-Essent)
	Laborelec
	IWPB (International Wood Pellet Buyers consortium)

[a]*PEFC* Programme for the Endorsement of Forest Certification Schemes, *SFI* Sustainable Forestry Initiative, *CSA* Canadian Standards Association, *ATFS* American Tree Farm System

from soy (mostly from Argentina or North America), biodiesel from palm oil (mostly from southeast Asia) and wood pellets (mostly from North America and Russia).

Evaluated schemes are listed in Table 6.1.

Governance and stakeholder involvement are crucial to ensure that certification schemes gain acceptance by the wide variety of stakeholders concerned with the sustainability of bioenergy. Most schemes are developed through a multi-stakeholder process, and are governed by a Board of Members, which (at first sight) equally represents all stakeholder groups. Although the general approach of these initiatives is very similar, the schemes differ in the way specific issues are dealt with and how they operate.

Chain-of-Custody systems All the sustainability certification initiatives have developed a Chain-of-Custody (CoC) standard, or intend to develop one (i.e. IWPB), but differ in which methodology should be applied and which parts of the chain are covered by the CoC certificate. All schemes provide procedures and guidelines on the specific requirements to comply with the CoC standards. Some schemes outline specific requirements for different actors within the supply chain. The physical segregation system[8] and the mass balance[9] system are the most commonly used

[8] Certified products are physically segregated from non-certified products at every facility along the supply chain.

[9] The amount of certified product sourced and sold by each supply chain actor is tracked. However, the certified product and associated documentation do not need to be sold together. The certified product can either be segregated (site level or tank level mass balance) or not (company level mass balance).

CoC systems. These are the traceability systems that are regarded as less prone to error and favoured by regulators because they provide direct incentives for fuel providers to ensure that the fuels they purchase and deliver meet sustainability requirements (Dehue et al. 2008).

All general biofuel (ISCC, RSB, 2BSvs) and crop-specific schemes (except for Bonsucro) refer to the tracking of sustainably produced products along the whole supply chain. All other certification schemes have partial CoC systems, excluding the farmer or biomass production and only include the operators handling or processing the certified product (wood products in the case of GGL, Laborelec, FSC and PEFC, and agricultural products in the case of GlobalGAP and SAN/RA).

Information handling The CoC tracking is based on continuous information about each stage of the supply chain taken by products from primary production at the forest, farm or crop site to the final user. It includes each stage of processing, conversion, transformation, manufacturing, trading and distribution where progress to the next stage of the supply chain involves an exchange of legal and/or physical control to ensure transparent transfer and traceability of certified feedstock/biofuel. In general, this information includes the volume, source of feedstock, type of feedstock and applicable certification number, together with sustainability data. Most CoC systems focus on the sustainability of feedstock production, and for biofuels all GHG emissions along the entire production chain must be included. Information on certificates and sustainability characteristics is generally transferred via online/electronic systems (i.e. 2BSvs, ISCC, RSB, RSPO) or through product declaration documents that are passed to the next operator in the supply chain (e.g. Bonsucro).

Assessment procedures In all schemes, each participating economic operator must be certified by a regularly accredited certifying body, and is subjected to an annual audit by an (independent) third-party auditor. Audit procedures appear similar in their intent, but there may be significant differences depending on the role that self-assessment, desk audits and field visits play, respectively. In certain cases of multi-site or group certification, only a sample of the entities involved in the certification are visited to verify that all conditions are met. The duration of most certificates varies from 1 to 5 years, after which the operation must be fully re-certified.

Recognition Standards often apply to similar or overlapping sectors, and for producers simultaneous certification according to more schemes, or recognition by multiple standards at the same time, can be an advantage. The costs of going through multiple audits can often be prohibitive for producers whose resources are limited. Thus, many standards have begun exploring ways to coordinate certification, thereby reducing the economic and administrative burden for economic operators. Improved consistency and collaboration for standards that are overlapping in either content or functions can lead to increased efficiency for standards themselves, and it can help scale up the use of certification generally, by making standards more available. Many schemes are recognized by other schemes or EU Member States. It is noteworthy to mention, for example, that many forestry and agricultural schemes are accepted by biofuel/bioenergy schemes as proof of sustainable wood production and agricultural biomass production. Recognition by other schemes, and especially by governments or the European Commission contributes to the credibility and assurance of a scheme. Mind that there is a risk of downgrading of sustainability/less transparency when certificates are recognized by other schemes along the chain (van Dam et al. 2012).

In general, it can be concluded that the voluntary sustainability schemes we examined tend to bring more credibility, accountability and transparency to the supply chain. They all address sustainability issues although they differ in the way these issues are addressed, e.g. differences in criteria and indicators, methodologies, audits, and level of transparency used. However, it must be noted that this complexity may create marketplace confusion and trade barriers.

6.4 Market Impacts

The existing bioenergy markets and trade are largely influenced by market characteristics and public policies. The market is shaped by a diverse group of factors such as resource availability and feedstock prices. These market elements are intertwined with intervention of a variety of national and regional policies, weaving a complex trading web. The implementation of sustainability requirements may have significant impact on the existing market and trade dynamics. This considerable complexity suggests a need to gain more insight into the interrelation between this wide range of factors and trading patterns to investigate the impact of sustainability requirements and certification.

In principle, there can be multiple effects of sustainability requirements on biomass production, availability and supply and trade, including (i) certain producing areas or resources can become excluded from specific markets (which can in turn enhance opportunities and market access for other potential suppliers), (ii) costs of production and feedstock supplies may increase, and (iii) certification can act to increase coherence along the supply chain and facilitate the realization of benefits (both ecological and socio-economic) associated with increased market access. Such mechanisms have been described for a few regions and resources (e.g. Smeets and Faaij 2010). Changes in trade flows are of particular interest when it comes to international (and intercontinental) bioenergy trade. The effectiveness of sustainability requirements may be undermined by leakage effects, i.e., producers decide to target new markets with less stringent requirements instead of improving their operations so as to comply with sustainability requirements on the markets where they have been present.

6.4.1 Trade Dynamics for Liquid Biofuels

The trade dynamics of liquid biofuels and solid biomass are significantly different. The liquid biofuels markets are rather mature markets and are closely related to agriculture commodities. Therefore, the markets are highly dynamic and complex and the actual impact of sustainability governance is not obvious. The liquid biofuels market is largely influenced by feedstock prices, which are closely related to food and feed commodities market, as well as crude oil prices (Lamers et al. 2011). For most of the crops, weather has been the determining factor for the supply, and hence the feedstock prices. Sustainability governance has reduced the size of certain supply chains, such as Argentinean soy-based biodiesel (SME) and southeast Asian palm oil based biodiesel (PME), especially since 2011, as only biofuels certified as sustainable are now accepted in the EU. This is mainly caused by the default GHG saving values set by the EC, which are below threshold for some SME and PME supply chains. So the influence of sustainability governance on these specific biofuels has been significant. However, to date, it has not affected the overall supply of sustainable biofuels, as fuels which fall under the double counting mechanism in the EU (such as waste-based biofuels) have increasingly dominated the market.

Additionally, the US has also developed a parallel market that effectively captures the Brazilian ethanol with a price premium (although Brazil's sugar-based ethanol production in 2011–2012 was more costly than US corn-based domestic supply). For this reason, obtaining sustainability certification to access the EU market has not been a priority for Brazilian producers. Brazil itself is currently facing a shortage of ethanol due to drought and poor investment in its cane belt (Reuters 2012). However, the ethanol trade between Brazil and the EU might recover in the near future, and the Brazilian Government, together with the private sector is fully engaged on the discussions for ethanol certification for European market (Dornelles, Brazilian ethanol exports and certification. Brazilian Ministry of Mining and Energy, personal communication, February 2013). We conclude that overall, for liquid biofuels, at the current mandate level, other factors have outweighed the sustainability governance in affecting trade dynamics, namely feedstock prices and local economic realities in individual markets. However, the influence from sustainability governance most likely will grow with the mandate level in the near future. The recent EC proposal to put a 5 % limit on food based biofuels (in an effort to address iLUC concerns) may depress the food crop-based biofuel trade and have a major impact on trade flows.

6.4.2 Trade Dynamics for Solid Biomass

The market is less complex and trade dynamics are more straightforward for solid biomass for energy, in particular wood pellets. The main market is the EU, and the primary drivers of development are national support policies, mainly for the promotion of renewable electricity production. Wood pellets are more expensive than coal, and this is not likely to change in the short term. Government subsidies determine the demand for solid biomass for energy, and subsidies typically come with sustainability requirements. It is still too early to make any conclusions about the effects of new sustainability requirements within, for example, the UK and the Netherlands, as utilities are still reacting to the policies. It is also important to consider that most wood pellet procurement strategies involve long-term contracts. Therefore, trade flows are unlikely to change on short notice. There is also a tendency for utilities to carry out vertical integration[10] for solid biomass operations.

Due to the nature of the market, solid biomass consumers, in particular wood pellet buyers, are cooperating to harmonize the existing certification schemes and systems (cfr. the IWPB initiative). Beyond sustainability considerations, harmonization of technical aspects and quality specifications is also one important consideration which requires coordination and harmonization. By putting effort in integrating diverse existing systems and regulations requirements, these actors aim to create a commodity market for solid biomass.

Due to the vertical integration and harmonization effort, sustainability certification is less likely to become a trade barrier for solid biomass in the future. Some sourcing areas might be excluded due to sustainability considerations in the processing section of supply chain rather than the harvesting. For example, Russian pellets were not accepted by the Dutch and Belgian utilities due to the use of natural gas for drying, which lowered the overall GHG savings.

The other important consideration would be the logistics issue (considering the emissions created through the transportation of solid biomass). However, trade conflicts in terms of solid biomass are different from liquid biofuels, such as the import of ethanol under different CN codes to get a lower tariff. Finally, the possible introduction of sustainability criteria on an EU level may be a major factor influencing solid biomass trade flows. Especially if strict thresholds for GHG emission reductions are introduced, or strict definitions of primary forests are introduced, a number of currently exporting regions such as Canada and Russia could be affected. Considering the current developments, we judge that the solid biomass market will likely continue to grow without dramatic changes in trade flows, but demand highly relies on government policies.

6.5 Issues Related to Sustainability Governance

6.5.1 Policies and Regulations

Need for long-term policies The developments in biofuels markets show clearly that uncertainty and ongoing changes in policies and regulations cause markets to stagnate. Prominent examples include the uncertainties about sustainability criteria for solid biomass in EU, the lengthy ongoing debates over iLUC risks for biofuels, and uncertainty about future policy supporting advanced biofuel mandates in the US. It should be kept in mind that stakeholders are making investment decisions

[10] Vertical integration means that energy producers try to control certain parts of the supply chain, e.g. through investments in plantations or pellet facilities.

now which establish long-term contracts, whereas governments may evaluate their policy year by year. Sustainability requirements are evolving and discussions on topics like iLUC for biofuels or carbon accounting for solid biomass are creating high uncertainties for companies, which in the future may need to comply with sustainability requirements that are unknown today.

In order to move the bioenergy sector as a whole towards more sustainable practices a legislative system that provides certainty over time is needed. A long-term policy strategy is considered an important driver to improve performances by defining clear objectives and creating a system of incentives (e.g. tax reliefs, subsidies). For their part, the regulations should lay down requirements which add credibility and encourage the development of transparent and comparable systems that are used by all stakeholders. Requirements which are costly and time-consuming and offer little added value (or reflect basic requirements covered by other regulations) should be avoided. Complications and restrictions can make tracing chain-of-custody too costly or create trade barriers for "certified" products. When changes need to be implemented because of new insights, these should be implemented through a transparent and incremental approach to avoid and minimize shock and market instability.

Different country approaches Policies and requirements differ from one country/region to the other due to different regional/country priorities, problems, government structures and processes. While sustainability criteria for biofuels (for transport) and bioliquids (for stationary energy) in the EU are directly related to the RED requirements and valid on EU-wide level, there are currently no obligated criteria for solid biomass on EU level. The main importing countries of solid biomass (UK, Netherlands, Belgium) have started to develop their own national sustainability requirements. At the same time, industrial and market business-to-business schemes are being developed. This has led and will lead to certification and verification schemes (voluntary and mandatory) that are not necessarily complementary or compatible. From a market/trade (and probably also policy) perspective, it may be preferable and more efficient to have a more aligned approach, possibly through a common international framework of (minimum) standards. This may not only lead to more international coherence and address the current proliferation of country/regional specific policies and requirements, but may also encourage the further internationalisation/globalisation of biomass/biofuel/bioenergy certification. More coordination would likely also make the interaction with the scientific community more effective, since scientists would not need to participate in numerous committees that essentially handle the same issues.

Discrimination between different end uses of the biomass and leakage issues Biomass for energy can be produced from various crops, which can also be used for food, feed or materials production. Currently only the use for biofuels needs to fulfil sustainability requirements on EU and US level. Similar commodities with similar environmental, social and GHG impact do not need to fulfil such requirements. Stakeholders producing biomass either for biofuels, for stationary energy, or for other applications (food, materials), are currently facing discrimination in conditions for being allowed to deliver their biomass. Farmers delivering corn to a transport biofuel installation need to be in compliance with the obligated sustainability criteria. The same farmer providing his corn to a biogas installation (combined with electricity production) doesn't need to fulfil these criteria, nor when he delivers his product to the food and feed markets.

An important issue is the willingness and cooperation of the biomass producers, especially from agriculture (for biofuels) and forestry (for solid biomass). If additional auditing is needed for agricultural products going to biofuels (as compared to other agricultural markets), or for solid biomass used for energy (as compared to the wood material market), this may diminish the willingness of the agricultural and forestry sector to deliver feedstock for bioenergy markets, unless there is a price premium for these certified products, which is hardly the case currently. Furthermore, diverting products with guaranteed sustainability to energy markets may not yield the intended sustainability benefits if it leads to displacement (leakage effects), when at the same time non-sustainable products supply the markets from which these sustainable products were diverted.

Criteria for sustainable production of liquid, solid and gaseous biomass should ideally be based on common concepts and be applicable to all uses of biomass. The producers of raw materials (such as forestry products, grains and oil seeds) do not necessarily know what the end uses will involve and most agricultural and forest commodities are processed into many different co-products. If everyone applied consistent sustainable land management criteria, risks of potential indirect effects and displacement (leakage) could be minimized. Sustainability criteria should be implemented in a practical way, bearing in mind that (a) practices should strive to permit sustained productivity from natural sources, (b) criteria should promote continual improvement toward meeting multiple goals of society (environmental, social and economic), and (c) these goals and priorities will change in response to changing needs, climate and other contextual conditions.

Need for common language In order to be consistent and transparent, including on a cross-sector level, there is a need for global/common definitions and processes on how the sustainability concept should be translated into practice, i.e. how to measure and weight sustainability dimensions and which criteria and indicators should be included. It is therefore very important to find a common language on "what is sustainable and how should sustainability be verified and documented", and using systems based on the same terminology.

A global initiative is needed to work towards global governance of land use principles and guidelines (e.g. a Multilateral Environmental Agreement) and to define a common language regarding implementation and verification. A uniform approach could gain credibility, acceptance and market penetration, and might be able to avoid different verification outcomes. This approach would allow for more efficient structures, save costs due to better management practices, ease administration tasks involved and make it unnecessary for industrial initiatives to create new standards. Costs derived of being part of a broader effort could be offset by access to much greater markets.

6.5.2 Voluntary Certification and Their Link to Policy and Legislation

Complimentary to policy Voluntary certification systems have become an important element in the mix of public policies and corporate strategies to promote the sustainable production of biomass. However, they will not be sufficient on their own. The history of forest certification has shown that it is unlikely that voluntary certification will be able to stop the production and use of non-sustainable biomass. Furthermore, other forms of governance, including legislation, are needed to address concerns that require regulation of resource use on larger scales, such as watershed planning, ecological zoning and rural development plans balancing nature conservation and socioeconomic development objectives. Also aspects such as iLUC and landscape-level carbon balances may be better addressed based on other instruments. For instance, on the longer term a global GHG emissions cap that regulates both fossil and biospheric carbon emissions could be one option providing flexibility. Countries may then decide to use a certain share of their permitted emission space to develop a bioenergy industry (resulting in some level of LUC emissions) to secure long-term domestic energy supply, or to generate export revenues.

Certification systems should not "try to do everything" and should be designed to effectively interact with other governance systems. Generally, legislation is intended to be (and needs to be) simple to apply, and it should sit at a relatively high level (i.e. create uniform regulations that can be applied at a national or international scale). Certification may serve as an on-the-ground tool that enables all actors involved in the supply chain to show compliance with legislative requirements and goals and create market incentives that recognize top performers. Additionally, these systems can decrease the administrative burden on governments by supporting the monitoring and control of implementation. Voluntary certification schemes generally are more adaptable and flexible than regulatory initiatives. Many of them revise their standards regularly. The International Organisation for Standardization (ISO), for example, reviews ISO standards every 5th year. Certification schemes can thus serve as innovative bodies to explore how sustainability requirements can promote improved performance over time by taking into account continuous scientific development and improvement of practices and based on this revise requirement levels in dialogue with stakeholders. They should complement regulations to improve awareness, facilitate discussion about the implications of certification and provide a forum for sharing information among stakeholders. However, consistent legislation and regulations supported by internationally agreed standards need to be implemented in conjunction to reach scale and create unified protection across systems, regions and countries, and reduce concerns of leakage.

Regional approaches When looking at the regional and international level, it is clear that some regions – in particular Europe and North America – already have a wide range of policies (legislation, regulations and guidelines) as well as implementation and control mechanisms in place to reasonably safeguard sustainable biomass production and regulate regional markets (although this is being challenged, sometimes rightfully so, by environmental groups). Here we are specifically referring to regulations that apply to bioenergy, forestry and agricultural management practices and other complementary regulations such as nature and environment protection regulations, land use and related planning acts.

Problems can arise in countries that lack strong, effective governance structures (i.e. lack of enforcement and control mechanisms) or where corruption is a problem. In these countries, context-specific approaches are needed to reduce the potential negative impacts of increasing the production and use of biomass. Chain of custody certification schemes offer a potential tool for improving the sustainability of biomass production as these systems include requirements designed to improve environmental and social practices and require regular third-party auditing and verification, and are able to operate across borders.

Capacity building Many developing countries are lagging behind with regard to implementation of sustainability governance because of financial, institutional and technical capacity. The implementation of sustainability systems – as conceived by developed countries – generally will require a much bigger leap for developing countries to reach a critical threshold because of the lack of technology and capital. Such requirements for data, analysis, technology or systems that are available in some nations, but not others, could create "non-tariff barriers" to international trade. The experience from forestry has shown that the adoption of certification schemes like FSC in developing countries can take decades. Based on the experiences of certification and sustainable management of resources in developed countries, it will be important to share information and technology and support capacity building to permit developing countries to participate productively in expanding certified global markets for sustainable bioenergy. As an example, the Global Bioenergy Partnership's Working Group on Capacity Building for Sustainable Bioenergy is fostering sustainable biomass and biofuels development and deployment, particularly in developing countries where biomass use is prevalent.[11]

6.5.3 Development and Implementation of Certification Schemes

Proliferation of schemes The proliferation of certification schemes in the past years has led to competition between different schemes. This may lead to improvement in the development of standards and tools for verification and monitoring. It may also provide insight into the relative effectiveness of different schemes for sustainability certification (design, implementation constraints, cost-benefits) as

[11] See http://www.globalbioenergy.org/programmeofwork/working-group-on-capacity-building-for-sustainable-bioenergy/en/ and a framework agreed to by GBEP participants on indicators to guide and measure the government programs and policies in the development of biomass and bioenergy http://www.globalbioenergy.org/fileadmin/user_upload/gbep/docs/Indicators/The_GBEP_Sustainability_Indicators_for_Bioenergy_FINAL.pdf

well as operational experience. The experience gained in developing schemes could also help to explore alternative models to meet sustainability goals.

On the other hand, the variety of sustainability initiatives and standards –with current lack of coherence and transparency, and considerable overlaps– may lead to confusion, lack of confidence and acceptance among the stakeholders including consumers. This may limit the effectiveness of sustainability governance, lead to a loss of clarity about the purpose, reduce participation, and cause distortion of the market. The risk is also that companies aim to use the commercially cheapest and least demanding certification scheme. 'Greenwashing' could undermine the credibility of schemes in general. With regard to the ease of scheme implementation, a good balance is needed between complexity and accessibility. If too many or overlying complex indicators are implemented, the certification process becomes too demanding, costly and difficult to manage and thus not attractive for users. Too little detail will lead to different interpretation of the principles and will raise doubts about the ability of the scheme to assure that the product and process meet the requirements of the scheme.

Consistency and recognition The main aim in the long term should be that terminology, definitions and methods converge to permit more consistency and transparency. Transboundary and trans-sector recognition would enable companies to expand market coverage without extra certification and related administrative and cost restraints. There are two types of recognition: (i) mutual recognition in the case of schemes that include the same or similar requirements (up to some level) and are implemented in an equal manner, and (ii) unilateral recognition in the case of schemes that complement each other (e.g. focus on different types of feedstock, parts of the chain and/or regions). In this way, stakeholders are not confronted with a multitude of audits and requirements depending on the type of schemes used along the supply chain or the end-use. For example, forestry or agricultural schemes could adapt to provide the necessary information required by other schemes for chain assessment, e.g. in terms of GHG emissions or land use change, or different schemes would be able to use the same chain of custody approach. There is already some movement towards recognition. Forestry schemes are accepted by ISCC; PEFC endorses numerous schemes globally. Also, RSB is in the process of recognizing other schemes. The agricultural scheme SAN by the Rainforest Alliance was benchmarked against RSB standards, and recognized by RSB as meeting them. It is important to state that this is only done when requirements are really similar between systems.

There is a need for more consistency in tools, models and guidelines used for implementation and verification. This would ensure that companies being certified are not evaluated in a manner that leads to different results for the same issue depending on the scheme or certification body. Many schemes have comparable objectives and common requirements regarding the design and setting up of infrastructure to manage these schemes. The sectors should build on experiences from forest and agricultural certification that have decades of experience in dealing with such problems.

Administrative burden and costs Certification places large demands on documentation and administrative procedures that can be costly and in particular may present a barrier for smallholders. Some schemes already allow group or stepwise certification as a way to gradually meet the requirements and make the investments needed to meet these requirements. This way it is possible to improve towards full compliance in a pace manageable for the producers. Governments could help in promoting and initiating group certification and lowering the administrative complexity to engage more smallholders in certification, and also control that there is sufficient momentum towards full compliance.

The administrative burden can increase if different schemes are used in the same supply chain, covering particular parts of the supply chain. To alleviate this barrier, coordination and recognition (unilateral and mutual) can be a vital measure to reduce administrative requirements, as was stated before.

6.6 Conclusions

International trade is an important component in the rapidly expanding bioenergy products markets. Some countries can rely on local resources, but many rely on imports due to insufficient local supply. As the main driver for bioenergy deployment is to enable society to transform to more sustainable fuel and energy production systems, sustainability safeguards are needed, either through binding regulations and/or voluntary systems, both for domestic and imported biomass.

There is currently a high number of initiatives and a proliferation of schemes. Markets would gain from more harmonization and cross-compliance. A common language is needed as 'sustainability' of biomass involves different policy arenas and legal settings. Standardization has proven to be very important (in other sectors) to create transparent markets and thereby facilitate rational production and trade.

Design of sustainability assurance systems (both through binding regulations and voluntary certification) should take into account how markets work, in relation to different biomass applications (avoiding discrimination among end-uses and users). It should also take into account the way investment decisions are taken, administrative requirements for smallholders, and the position of developing countries. Policy pathways should be clear and predictable, and future revisions of sustainability requirements should be open and transparent.

References

Chum, H., et al. (2012). 2011: Bioenergy. In O. Edenhofer et al. (Eds.), *IPCC special report on renewable energy sources and climate change mitigation*. Cambridge/New York: Cambridge University Press.

Dehue, B., Hamelinck, C., Reece, G., de Lint, S., Archer, R., & Garcia, E. (2008). *Sustainability reporting within the RTFO: Framework report*. Ecofys BV, Jan 2008. Available at: http://biomass.ucdavis.edu/secure/materials/sustainability%20committee/Ecofys%20sustainability-reporting%20May07.pdf. Accessed Feb 2013.

EC. (2009). *Directive 2009/28/EC of the European Parliament and of the Council of 23 April 2009 on the promotion of the use of energy from renewable sources and amending and subsequently repealing Directives 2001/77/EC and 2003/30/EC. Apr 2009*. Available at: http://eur-lex.europa.eu/LexUriServ/LexUriServ.do?uri=OJ:L:2009:140:0016:0062:EN:PDF. Accessed Feb 2013.

EC. (2010). *Report from the Commission to the council and the European Parliament on sustainability requirements for the use of solid and gaseous biomass sources in electricity, heating and cooling. COM(2010)11, Feb 2010*. Available at: http://ec.europa.eu/energy/renewables/transparency_platform/doc/2010_report/com_2010_0011_3_report.pdf. Accessed Feb 2013.

EC. (2012). *Proposal for a Directive of the European Parliament and of the Council amending Directive 98/70/EC relating to the quality of petrol and diesel fuels and amending Directive 2009/28/EC on the promotion of the use of energy from renewable sources. COM(2012)595, Oct 2012*. Available at: http://ec.europa.eu/clima/policies/transport/fuel/docs/com_2012_595_en.pdf. Accessed Feb 2013.

Goh, C. S. et al. (2013). *Impacts of sustainability certification on bioenergy markets and trade*. Subreport (Task 3) of the study 'Monitoring Sustainability Certification of Bioenergy', joint study involving IEA Bioenergy Task 40, Task 43 and Task 38. Available at: http://bioenergytrade.org/downloads/iea-sust-cert-task-3-final2013.pdf. Accessed Feb 2013.

Goovaerts, L. et al. (2013). *Examining sustainability certification of bioenergy*. Subreport (Task 1) of the study 'Monitoring Sustainability Certification of Bioenergy', joint study involving IEA Bioenergy Task 40, Task 43 and Task 38. Available at: http://bioenergytrade.org/downloads/iea-sust-cert-task-1-final2013.pdf. Accessed Feb 2013.

ISCC. (2012). Available at: http://www.iscc-system.org. Accessed 13 June 2012.

IWPB. (2012). *Initiatives wood pellet buyers*. Available at: http://www.laborelec.be/ENG/initiative-wood-pellet-buyers-iwpb. Accessed Feb 2013.

Lamers, P., Hamelinck, C., Junginger, M., & Faaij, A. (2011). International bioenergy trade – A review of past developments in the liquid biofuel market. *Renewable and Sustainable Energy Reviews, 15*, 2655–2676.

Lieback, J., & Kapsa, K. (2011, June 6–10). *Sustainable biomass in the light of certification and assessment schemes*. Presented at the 19th European Biomass Conference and Exhibition, Berlin. Available at: http://www.etaflorence.it/proceedings/?detail=6891. Accessed Feb 2013.

NTA 8080. (2009). *Dutch Technical Agreement, NTA 8080*, Sustainability criteria for biomass for energy purposes. Available at: http://www.sustainable-biomass.org/publicaties/3938. Accessed Feb 2013.

O'Connell, D., Braid, A., Raison, J., Handberg, K., Cowie, A., Rodriguez, L., & George, B. (2009). *Sustainable production of bioenergy: A review of global bioenergy sustainability frameworks and assessment systems*. Rural Industries Research and Development Corporation, Canberra. Available at: https://rirdc.infoservices.com.au/items/09-167. Accessed Feb 2013.

Pelkmans, L., Devriendt, N., Goovaerts, L., & Schouwenberg, P. P. (2012). *Implementation of sustainability requirements for biofuels and bioenergy and related issues for markets and trade*. Study accomplished within IEA Bioenergy Task 40. Available at: http://bioenergytrade.org/downloads/t40_implsustcert_final-report_march-2012.pdf. Accessed Feb 2013.

Pelkmans, L. et al. (2013). *Recommendations for improvement of sustainability certified markets*. Subreport (Task 4) of the study 'Monitoring Sustainability Certification of Bioenergy', joint study involving IEA Bioenergy Task 40, Task 43 and Task 38. Available at: http://bioenergytrade.org/downloads/iea-sust-cert-task-4-final2013.pdf. Accessed Feb 2013.

Reuters. (2012). *Analysis: Brazil ethanol returns to US as biofuel rules pave way*. Available at: http://www.reuters.com/article/2012/09/20/us-ethanol-brazil-exports-idUSBRE88J14J20120920

Scarlat, N., & Dallemand, J. F. (2011). Recent developments of biofuels/bioenergy sustainability certification: A global overview. *Energy Policy, 39*, 1630–1646.

Smeets, E., & Faaij, A. (2010, March). The impact of sustainability criteria on the costs and potentials of bioenergy production: Applied for case studies in Brazil and Ukraine. *Biomass & Bioenergy, 34*(3), 319–333.

Stupak, I., Lattimore, B., Titus, B. D., & Smith, C. T. (2011). Criteria and indicators for sustainable forest fuel production and harvesting: A review of current standards for sustainable forest management. *Biomass and Bioenergy, 35*(8), 3287–3308.

Stupak, I. et al. (2013). *Survey on governance and certification of sustainable biomass and bioenergy*. Subreport (Task 2) of the study 'Monitoring Sustainability Certification of Bioenergy', joint study involving IEA Bioenergy Task 40, Task 43 and Task 38. Feb 2013. Available at: http://bioenergytrade.org/downloads/iea-sust-cert-task-2-final2013.pdf. Accessed Feb 2013.

van Dam, J., et al. (2010). *Update: Initiatives in the field of biomass and bioenergy certification, Background document 'From the global efforts on certification of bioenergy towards an integrated approach based on sustainable land use planning'*. Available at http://www.bioenergytrade.org/downloads/overviewcertificationsystemsfinalapril2010.pdf. Accessed Feb 2013.

van Dam, J., Ugarte, S., & van Iersel, S. (2012). *Selecting a biomass certification system – a benchmark on level of assurance, costs and benefits*. Study carried out for NL Agency. Mar 2012. Available at: http://www.agentschapnl.nl/sites/default/files/bijlagen/Report_study_Selecting%20a%20Certification%20System%20March%202012.pdf. Accessed Feb 2013.

Chapter 7
Drivers and Barriers for Bioenergy Trade

Martin Junginger, Peter-Paul Schouwenberg, Lars Nikolaisen, and Onofre Andrade

Abstract There are several drivers responsible for the strong increase in biomass trade over the past decade: concerns regarding the effects of climate change remain unchanged, and policy targets for renewable energy for 2020 have so far remained (largely) intact despite the economic crisis. At the same time, the list of barriers potentially hampering the further growth is long and very heterogeneous. Import tariffs and anti-dumping measures have been the topic of dispute between the main producing and consuming regions of ethanol and biodiesel for the last decade, and also technical standards for biodiesel have been criticized, as they may put biodiesel made from soy and palm kernel oil at an disadvantage. For solid biomass, phytosanitary measures are one of the most important barriers preventing the trade of softwood wood chips for energy. Also health and safety issues related to transporting and storing solid biomass still need further attention. For bioenergy trade towards the EU to grow further, long-term investment security is required, a clear and stable sustainability framework has to be in place, and the legal and technical aspects of solid biomass have to be rapidly standardized. The current crisis is likely to influence the climate change business negatively in the short term, but under a stable regulatory framework, even if in short term profit is slow, companies with a long term vision would still find sustainable projects attractive enough to invest.

M. Junginger (✉)
Copernicus Institute, Utrecht University, Utrecht, The Netherlands
e-mail: h.m.junginger@uu.nl

P.-P. Schouwenberg
RWE Essent, s-Hertogenbosch, The Netherlands

L. Nikolaisen
Centre for Biomass & Biorefining, Danish Technological Institute, Taastrup, Denmark

O. Andrade
Argos Energies, Rotterdam, The Netherlands
e-mail: andrade.onofre@gmail.com

M. Junginger et al. (eds.), *International Bioenergy Trade: History, status & outlook on securing sustainable bioenergy supply, demand and markets*, Lecture Notes in Energy 17, DOI 10.1007/978-94-007-6982-3_7,

7.1 Introduction

As shown in amongst others Chaps. 2 and 3, the international trade of various bioenergy commodities has grown rapidly, driven mainly by climate and renewable energy support policies. However, this growth is also hampered by a number of barriers. For example, the export of palm oil from South East Asian countries for the production of renewable electricity or biodiesel in Western Europe has been heavily criticized by NGOs (see e.g. Zakaria et al. 2009). Also the possible impacts of other internationally-traded liquid biofuels (such as soy-based biodiesel and ethanol from corn) have received increasing public (and largely negative) attention. On the other hand, international bioenergy trade can also offer opportunities for economic growth and socio-economic development for exporting countries, and may enable countries with few domestic biomass resources to meet their renewable energy targets, gain more fuel diversity and improve security of supply. The role of international trade in (especially liquid) biofuels has been discussed by several authors (see e.g. Dufey 2007; EurActiv 2009; Heinimö and Junginger 2009; Londo et al. 2010; Murphy 2008; Oosterveer and Mol 2010; Steenblik 2007; Zarrilli 2008).

This chapter aims to provide an overview of both the most important drivers and barriers for bioenergy trade. It is largely based on a similar paper published in 2011 (Junginger et al. 2011). The chapter focuses mainly on three internationally-traded bioenergy commodities: bioethanol, biodiesel and wood pellets. The choice for these commodities is motivated by the strong growth of trade of these commodities in the past years (see also Chaps. 2 and 3).

In this chapter, we define 'barriers for international bioenergy trade' very broadly, mainly determined by what various stakeholders may perceive as a barrier to bioenergy trade. Principally, we define a bioenergy trade barrier as *any issue that either directly or indirectly hinders the growth of international trade of biomass commodities for energy end-use*. It is difficult to draw a clear line what (indirect) *trade* barriers are, and what *general* barriers hamper the use of biomass (irrespective of being traded or used domestically). For example, the current food-vs-fuel debate (e.g. should vegetable oils be used as feedstock for biodiesel) and the indirect land use debates affect biomass use in general, and is not discussed here as specific barrier to trade. Yet, this debate and resulting policy measures is likely to have direct impacts on the amount of ethanol, vegetable oils and biodiesel traded globally in the coming years. Also, the global economic crisis is affecting bioenergy trade, but as it also affects biomass production, consumption, oil prices etc., we do not list it as a 'trade barrier'. Likewise, also for drivers, we realize that drivers for bioenergy trade are often also drivers for bioenergy in general. Nevertheless, we point out if and how these drivers specifically support trade.

7.2 Drivers for Bioenergy Trade

Bioenergy, as (almost) all forms of renewable energy, has needed support in the past, in the form of regulatory, legal and economic policy support addressing mainly climate change and (related) the ambition to increase amount and share of renewable energy.

These drivers have also been in the majority of cases the main reason behind large-scale trade of liquid and solid bioenergy, possible exceptions being cross-border fuelwood and waste wood trade (Junginger et al. 2010). Other possible drivers, such as geopolitics and related energy-security concerns or rural development and the search for new agricultural commodities have so far been of minor or negligible importance, but may become more important in the coming decades.

7.2.1 Legal and Regulatory Drivers

7.2.1.1 Globally: Kyoto

The Kyoto Protocol to the United Nations Framework Convention on Climate Change (UNFCCC) sets binding obligations on industrialized countries to reduce emissions of greenhouse gases. The UNFCCC is an international environmental treaty with the goal of achieving the "stabilization of greenhouse gas concentrations in the atmosphere at a level that would prevent dangerous anthropogenic interference with the climate system". Since its adoption in December 1997 and its entering into force in February 2005, the Kyoto protocol has probably one of the largest (indirect) drivers for (trade in) bioenergy. For example the EU-ETS was established, which means that large industrial GHG emitters in the EU have to pay for fossil emissions – a direct economic incentive aiding the use of more biomass (see also Fig. 7.3). However, with the very limited progress achieved during the recent climate conferences in amongst others Copenhagen and Durban (including the failure to agree on any binding emission targets beyond 2020), the US, Canada and Australia not being part of the Kyoto protocol, and a historically low CO_2 price of below 5 €/tonne in early 2013, it remains questionable whether the Kyoto protocol (and climate change mitigation efforts in general) will remain a driver of significance. This becomes even more uncertain due to the latest developments in Brussels regarding the failed voting to take a substantial volume of CO_2-emission rights out of the market (back-loading). In practice, this will mean that the EU-ETS will no longer be a driver for the use of biomass. We now have to wait for national solutions to realize the EU targets in 2020–2030, if these GHG emission targets will remain in the first place. It is our view that biomass co-firing is one of the cheapest solution on the short term to realize the EU-targets. For the long term, other solutions will have, i.e. the transition towards a Biobased Economy, in which biomass increasingly will be used first for material application and only in a final stage used for energy purposes (cascading).

7.2.1.2 European Union: Renewable Energy Directive (Including Biofuels)

The Renewable Energy Directive (RED) is a European Union directive which mandates levels of renewable energy use within the European Union and was published in April 2009. The directive requires EU member states to produce a pre-agreed proportion of energy consumption from renewable sources such that the EU as a whole shall obtain at least 20 % of total energy consumption from renewables by 2020. As part of the RED, in 2010, all member states had to submit national renewable energy actions plans (NREAPS). From these NREAPS, it has become clear that solid, liquid and gaseous biomass is likely to play an important role to meet the renewable energy targets of many European countries. In many EU countries, the expected demand for biomass is also higher than domestic supply, which will necessitate biomass imports (see Fig. 7.1). Already in the past few years, several EU countries have increased their solid biomass and liquid biofuels imports (both from other EU countries and from outside the EU, see also Chaps. 2 and 3).

As part of the RED, EU member states are also required to achieve a 10 % share of renewable transport fuels in 2020. Towards the end of 2012, the European Commission proposed that a maximum of 5 % (i.e. half) for this target may be covered using biofuels derived from food crops. The other half will have to be sourced from alternative sources, such as used cooking oil, and 2nd and 3rd generation advanced biofuels, e.g. from lignocellulosic feedstocks and algae. How this change in policy will affect trade flows is yet to be seen (see also Sect. 7.3.5 of this chapter).

7.2.1.3 USA: RFS-2

Looking at the United States of America the RFS2 becomes more and more import for the international Bioenergy Trade. This Renewable Fuel Standard (RFS) – included in the Energy Independence and Security Act (EISA) – provides provisions on the promotion of biofuels (especially cellulosic biofuels). EISA mandates minimum GHG reductions from renewable fuels, discourages use of food and feed crops as feedstock, permits use of cultivated land and discourages (indirect) land-use changes. The RFS is a federal program that requires transportation fuel sold in the U.S. to contain a minimum volume of renewable fuels. The RFS originated with the Energy Policy Act of 2005 and was expanded and extended by the Energy Independence and Security Act of 2007 (EISA). The RFS program requires renewable fuel to be blended into transportation fuel in increasing amounts each year, escalating to 36 billion gallons by 2022. Each renewable fuel category in the RFS program must emit lower levels of greenhouse gases relative to the petroleum fuel it replaces. By 2022, the majority of the fuels consumed must be from cellulosic ethanol, biomass-based diesel and other advanced fuels (EPA 2013; AFDC 2013). This may also have consequences for international biofuel trade: As Brazilian sugarcane ethanol production qualifies as an “advanced” biofuel under the RFS greenhouse gas (GHG) calculations, and corn-based ethanol only qualifies as a “renewable” biofuel, it is quite possible that Brazilian ethanol will increasingly be imported to fulfill the advanced mandate (Irwin and Good 2012). For this reason, in 2012, the US imported significant amounts of Brazilian ethanol, despite the fact that this ethanol was more expensive than the domestically produced ethanol. Whether imports will further increase in years to come will mainly depend on the price development of other advanced biofuels, and the cost of biodiesel, and possible US policy developments, such as the reinstatement of the $1 per gallon tax credit for biodiesel blending (Irwin and Good 2012).

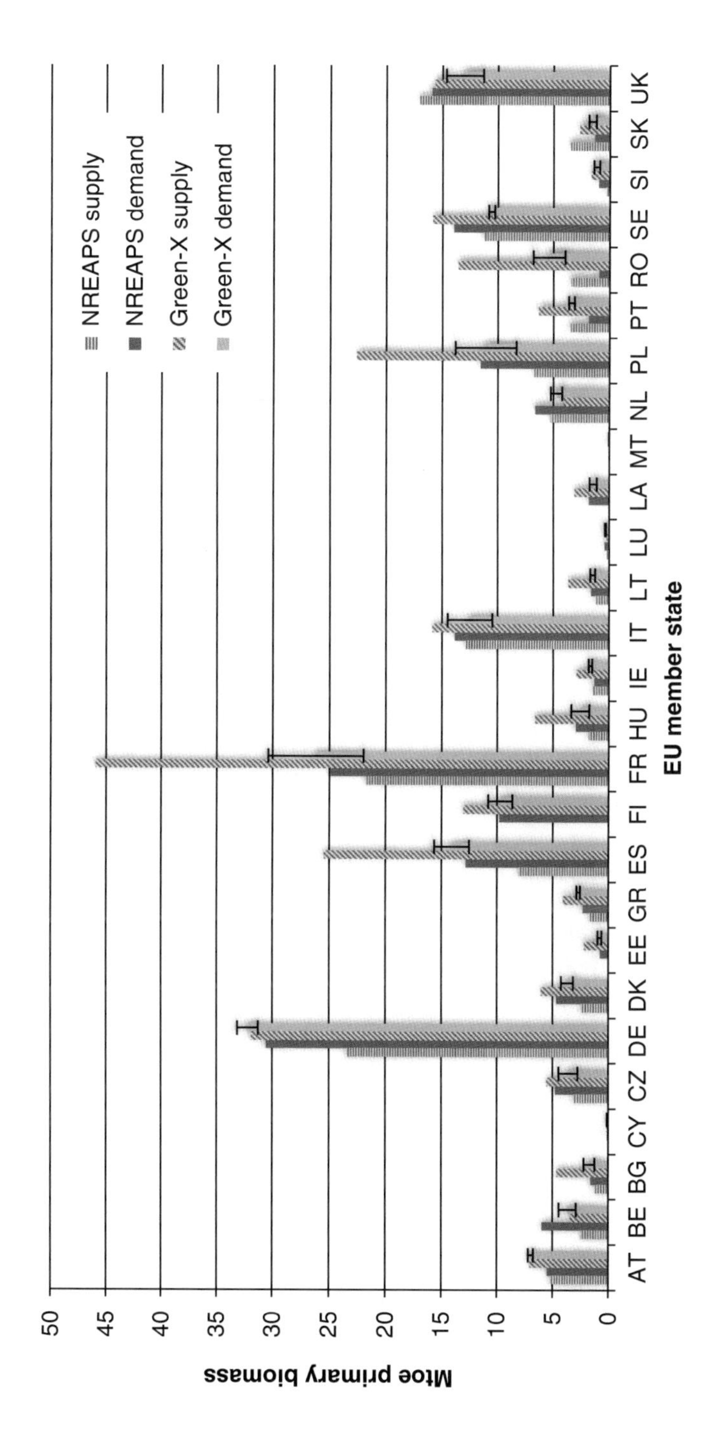

Fig. 7.1 Expected primary biomass supply and demand for electricity, heat and biofuels in 2020 based on the National Renewable Action Plants; and Green-X supply and demand ranges for the scenarios Business as Usual (*BAU*), BAU Mitigated barriers and Strengthened National Support for 2020 (Source: Hoefnagels et al. 2011)

7.2.2 Economic Drivers

The economic drivers of bioenergy can be split into subsidies and commodity prices. In this section, first, the subsidies are described which are directly linked to the legal and regulatory drivers. The legal and regulatory frameworks mentioned in the previous section are providing the basis for governments to implement economic policy support schemes in which they enable sustainable energy and fuel production. Second, this section will focus on the price level of conventional commodities and their impact on bioenergy. Finally we discuss how different biomass types could be develop to become a true commodity.

7.2.2.1 Subsidy Schemes

Renewable energy production is being subsidized mainly in two ways (based on the definitions provided by Wikipedia.org):

1. A **feed-in tariff** is an incentive structure to encourage the adoption of renewable energy through government legislation, by granting higher-than-market electricity prices to producers using renewable sources. The legislative assurance of higher prices over a predetermined period of time, helps overcome the cost disadvantages of renewable energy sources. The rate may differ among various forms of power generation. Feed-in tariff laws were in place in 46 jurisdictions across the world by 2007. The UK implemented a feed-in tariff by 2010 in addition to its renewable energy quota scheme (which utilizes Renewable Obligation Certificates – ROCs)
2. **Green certificates** are tradable commodities proving that certain electricity is generated using renewable energy sources. Typically one certificate represents generation of 1 MW hour of electricity. What is defined as "renewable" varies from one certificate trading scheme to another, but in any case, green certificates represent the environmental value of renewable energy generated. The certificates can be traded separately from the energy produced. Several countries use green certificates as mean to make the support of green electricity generation closer to market economy instead of more bureaucratic investment support and feed-in tariffs. Such national trading schemes are in use in e.g. Poland, Sweden, the UK, Italy, Belgium, and some US states. In all of these cases the supply of renewable energy is achieved by obliging generators to obtain a portion of their electricity from renewable energy sources or alternatively cover such portion by purchasing green certificates on the market (thus imposing an additional cost to "dirty electricity"). Hence a market is created in green electricity certificates which make renewable energy developers more competitive.

Figure 7.2 illustrates how the country demand for biomass generation driven by renewable energy subsidies defines the target third-party sales market. Note that the situation shown in Fig. 7.2 may change over time, as national governments adapt

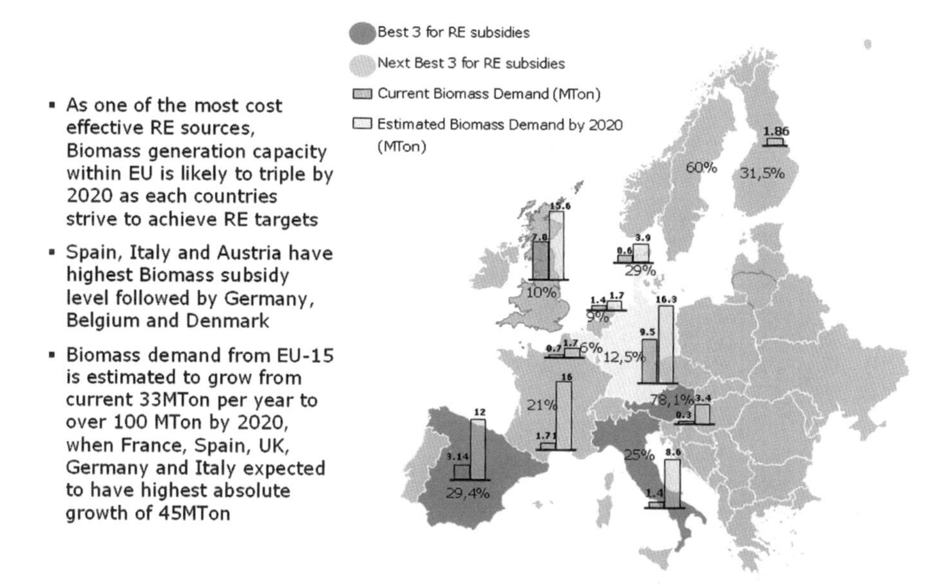

Fig. 7.2 Overview of EU countries with highest subsidy levels for bioenergy

their policies to meet the renewable energy targets set out in the NREAPS. As of early 2013, the UK is also likely to be amongst the top six countries in the EU to promote bioenergy. In the UK, the proportion of electricity generated by biomass in coal fired power stations is eligible for ROCs (Renewables Obligation Certificates), which have an intrinsic, market value and also avoids carbon emission allowances costs. All 16 major UK power plants are now co-firing a proportion of biomass, at an average level of 3 % (energy basis) making use of a range of fuels including wood (virgin and recycled), olive cake, palm kernel expeller, sewage sludge and energy crops (Biomass Energy Centre 2013).

7.2.3 Commodity Prices

The main drivers for bioenergy are subsidies and – not to be underestimated – commodity prices of the fuels it substitutes, and the price of CO_2. In Fig. 7.3, typical breakdowns of the cost and revenue structures of a co-firing plant and a standalone biomass power plant are represented. On the revenues side, power, carbon and avoided fuels (coal or gas) cost are prevailing. On the cost side, the biomass prices are dominant. Investment (i.e. fixed) and O&M (i.e. variable) cost only contribute a marginal share to the total costs. Note however that these figures are highly depending on the specific local situation, and should only be considered as an illustration (Fig. 7.3).

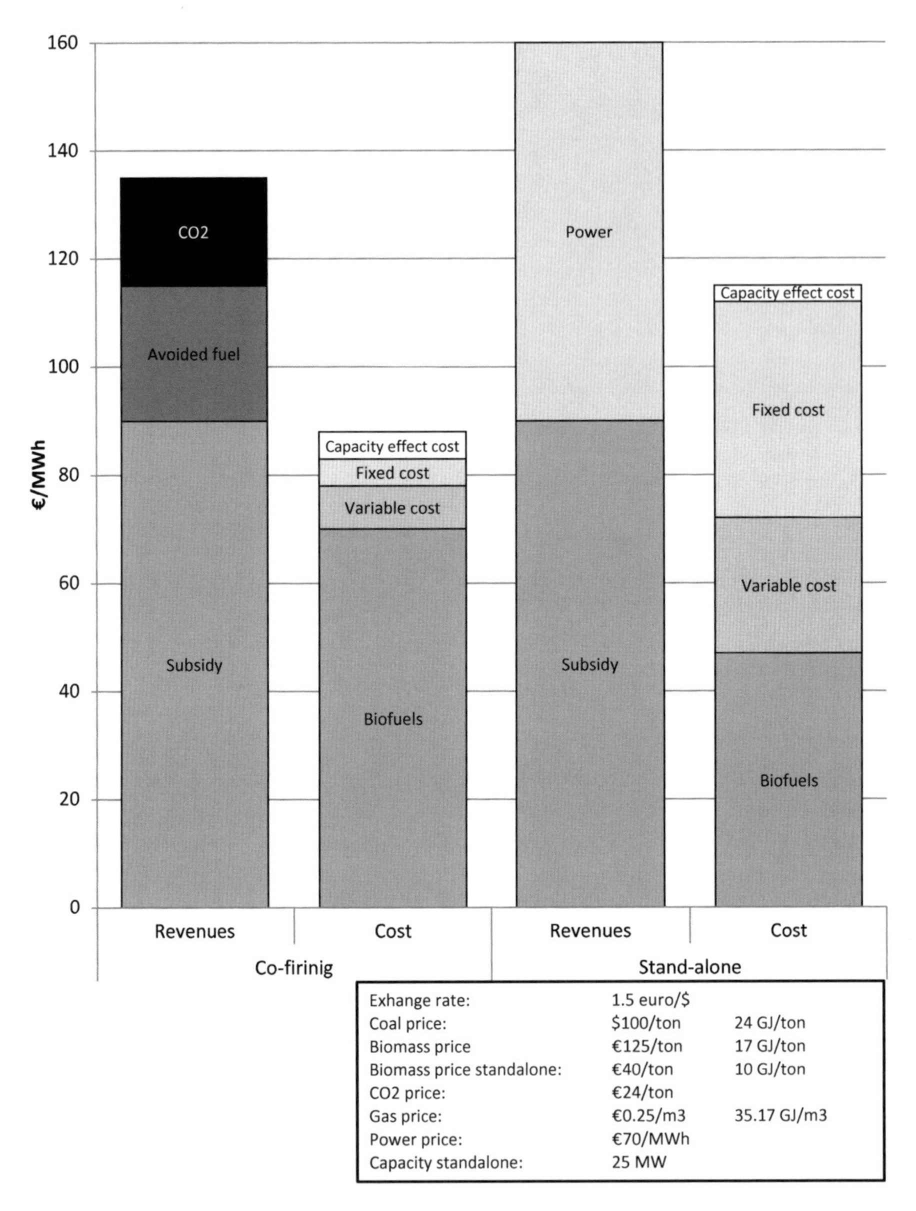

Fig. 7.3 Example of an illustrative revenue and cost structures for co-firing versus stand-alone biomass combustion for power production in Northwest Europe

The graph above is based on a situation of 90 €/MWh subsidies, and is a general presentation of the cost structures for plants in North-West Europe. The assumed fuel for co-firing is wood pellets, whereas for the stand-alone plant, forest residues are used (e.g. from local sourcing, from the Ardennes, from Germany or shipped in through ARA). Alternative fuels like palmstone shells and agricultural residues are also possible, but logistics challenges may be significant because of the because of the huge volumes required and long distances. Also clean, unpainted post-consumer wood from industrial sources might be possible, although currently, this could create a tension with the German waste wood market.

Going into more detail, the biofuels commodity prices are influenced by the biofuel supply chain elements, secondary commodity price effects:

- ***Coal prices/oil prices***: represent an economic variable determining choices on development of alternative investments in energy/electricity generation and enhance diversification of fuel/technology mix to improve stability on energy prices
- ***Wood pellets prices***: can be very volatile, because of production and demand figures, and also weather conditions. The industrial pellets compete also with the DIN+ pellets for the retail market. In principal the retail market can pay higher prices.
- ***Vegetable oil prices***: are driven by the food – and feed industry. These markets can pay higher prices than the Electricity markets. Furthermore the Electricity markets face a huge discussion concerning the sustainability of the oil. In principal power shouldn't compete with food and feed.
- ***CO_2-prices***: even though the CO_2 prices represent the key driver for emission projects and the biomass projects can be profitable even without the carbon rights, the possibility of obtaining these rights can only make this business opportunity even more attractive.
- ***Raw materials markets***: economic cycles in upstream industries connected to the renewable energy sector (e.g. paper and pulp companies providing raw materials to pellets plants).
- ***Commodity Freight/transport costs***: the biofuels and biomass transportation costs are currently high. In order to increase its attractiveness to the new investors in this sector, they need to be decreased either directly or through subsidies
- ***Currency exchange rates***: since the biofuels market is quoted in USD, the more this currency depreciates against EUR, the more attractive the purchase of biofuels becomes for the European generation projects.

Looking at historical and forward prices expectations (see Fig. 7.4), the steep peak that took place in 2008 was a one-time event in the past decade. It is rather unlikely that commodity prices will reach these levels again on the short term. Instead, it is more plausible that prices will slowly increase leveling at historical levels or slightly above them.

The raw material (wood) market will be a key market for primary biofuels such as chips and wood pellets. It is also possible that competition with other industries may take place. In the future cheap woody biomass feedstocks will be required also by the paper and pulp industry and the 2nd generation biofuels industry. As illustrated in Fig. 7.5, sourcing in the more expensive parts of the (wood) market will be necessary to provide continuity of supply. Currently, wood pellets producers source mainly bark, sawdust, wet- and dry woodchips. In order to secure the raw material supply in the (near) future, wood pellet production facilities are increasingly taking the price of small whole trees (typically pulp quality and -sized

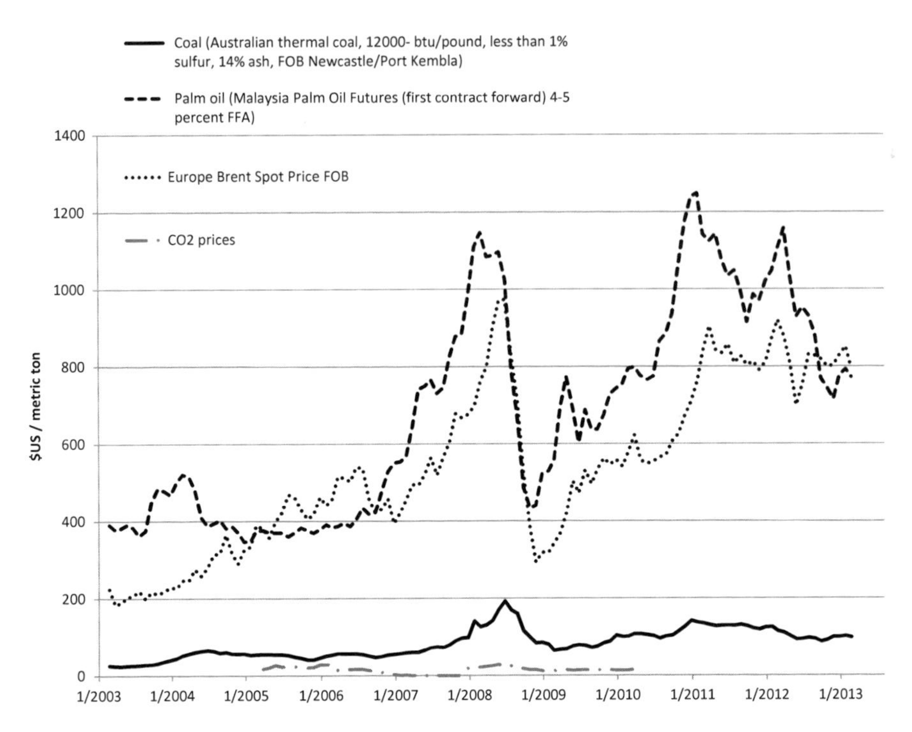

Fig. 7.4 Prices of fossil fuels and palm oil in the past decade (Source: Coal and palm oil prices – Indexmundi 2013; Brent prices – EIA 2013; CO_2 prices – Own data)

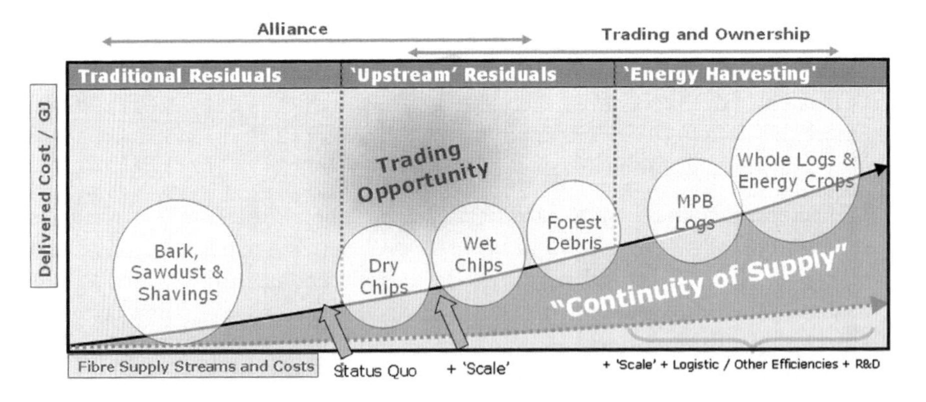

Fig. 7.5 Overview of woody biomass feedstocks types and prices

material) into consideration and invest in chipping equipment. This of course will increase their cost price. The increase of the cost of raw materials may in some cases be bearable for wood pellet mills, as typically part of the loans for the investment are repaid after year 3 or 4. Also, the price of woody biomass will also strongly depend on local availability of different woody feedstocks and demand

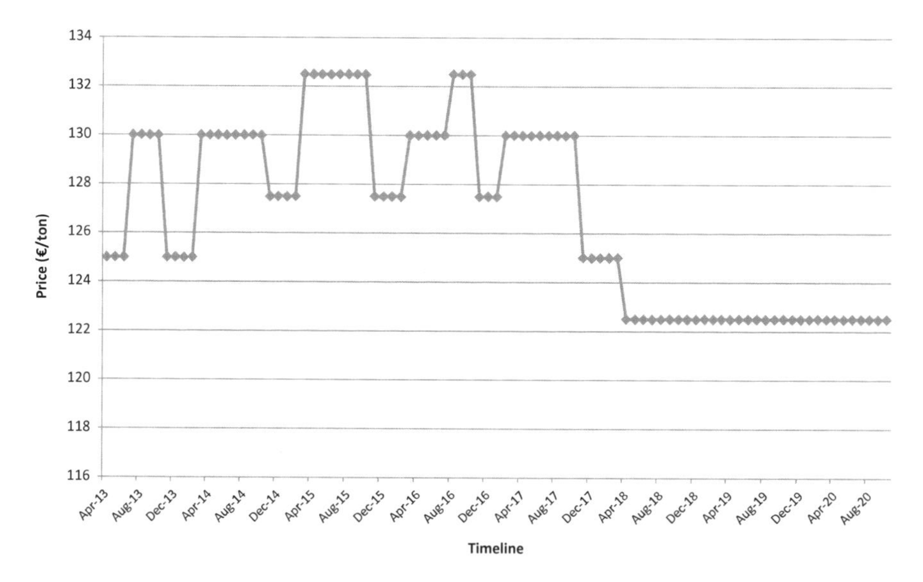

Fig. 7.6 Forward curve ET wood pellets, cost, insurance & freight (*CIF*) Rotterdam (based on authors own data)

by other industries. As an indication, the authors estimate that cost prices of raw woody materials in the near future will be around 25–30 €/tonne (dry) in Canada, 30–35 €/tonne (dry) in the US, and 50–65 €/tonne (dry) in Europe. Raw materials from landfill (nature conservation) have a discount of typically 10–15 €/tonne (dry).

The forward curve for wood pellets (delivered CIF to Rotterdam, see Fig. 7.6) is the same for all production areas. The difference in the costs of raw materials must be compensated with differences in logistic costs and/or less profit. The pricing in the summer is always lower, because in this period when there is less demand. The forward curve declines to about 123 €/tonne after 2018, as during this period it is assumed that sufficient production capacity will be in place and at the same time less subsidies may be available.

7.3 Barriers for Bioenergy Trade

7.3.1 National/Regional Protectionist Policies and Tariff Barriers

To mitigate the generally higher production costs of liquid biofuels, governments in many countries have supported domestic production and use through policy incentives, such as tax exemptions and subsidies. In some cases, these support schemes may shield domestic producers from foreign competition and thus limit international trade. These incentives are often geared towards the promotion of domestic agricultural feedstocks and interests, rather than the promotion of biofuels with economic, energetic or environmental advantages. In other cases, policy incentives can (directly or indirectly) support the export of biofuels. Importantly, support measures in developed countries have implications for developing country producers by reducing their competitiveness, creating global inequalities and distorting international trade. Interestingly, no such measures are known to the authors for solid biomass.

In principle, three different mechanisms can be distinguished:

The most important and well-known trade barrier are ***import tariffs***. Import tariffs have been used for decades by different countries for ethanol, and with the rise of biodiesel trade also for biodiesel. As of March 2012, the EU applies a € 0.192/l tariff on undenaturated ethanol while the import duty for denaturated ethanol is € 0.102/l. The tariffs do not distinguish between the different uses of ethanol (beverage, fuel, industrial). In March 2012, the European Commission also published a customs regulation for ethanol blends containing up to 30 % gasoline to a flat rate of € 0.102/l, which was previously taxed only 6.5 % ad valorem (or around € 0.035/l). The EU's decision contrasts with the abolishment of the ($ 0.142/l) import duty on ethanol charged by the US until the end of December 2010 (Kfouri 2012). Also, the EU in February 2013 announced it would impose a $ 0.0803/l tariff on U.S. ethanol imports for 5 years after November 2011 complaint that U.S. ethanol importers were selling the fuel below cost – or "dumping" – a practice that EU ethanol producers say caused ethanol prices in Europe to fall (Farm Futures 2013). However, the EU maintains several preferential trade arrangements with developing countries with either no duties or reduced tariffs for ethanol, including the Generalized System of Preferences (GSP, which applies to many developing countries), the Cotonou Agreement (African, Caribbean, and Pacific countries; or ACP Group), the Everything But Arms (EBA) initiative (for developing countries). In general, the most-favoured nation (MFN) tariffs range from roughly 0 to 50 % on an ad valorem equivalent basis in the OECD, and up to 186 % in the case of India (Steenblik 2007).

Also the export of biodiesel from the US to the EU has been the subject of fierce debate and import tariffs (also known as anti-dumping measures) in recent years. Over the course of 2007 ad 2008, a volumetric excise tax credit in the US led to massive exports of B99 biodiesel from the US to the EU (the so-called splash-and-dash effect). In 2009, the European Commission approved antidumping and anti-subsidy rights on American biodiesel imports between €213 and €409 per tonne for a period of 5 years (EurObserv'ER 2009), which was later-on extended to Canada to prevent triangular trade. In May 2013 the European Commission issued Regulation imposing a provisional anti-dumping duty on imports of biodiesel origination from Argentina and Indonesia. For biodiesel originating from Argentina, the provisional duty rate amounts between €65.24 and €104.92 per tonne net. For biodiesel originating from Indonesia, the provisional duty rate amounts between €24.99 and €83.84 per tonne net.

Second, also ***export tariffs*** can have an impact on trade of biomass commodities. For example, differential export taxes exist in Argentina and Indonesia, favoring the production and export of the finished product biodiesel rather than (soybean/palm) oil. It is difficult to quantify in how far these preferential export tariffs have de facto spurred production and export of biodiesel in both countries – however, the fact is that both countries continue to export biodiesel, while e.g. Malaysian biodiesel exports were very low in 2012. While such preferential tax systems actually increase traded volumes, producers in other countries may experience them as unfair competition. The EU anti-dumping measure applicable to biodiesel from Indonesia and Argentina is directly related to biodiesel export tariffs in those countries.

Third, measures to ***promote domestically produced biofuels over imported biofuels*** need to be mentioned. For example, it has been reported that biofuels produced in Belgium, France, Italy, Spain and several US states received policy support such as tax exemptions or volumetric subsidies on locally-produced biofuels (Euractiv 2009; Steenblik 2007; Koplow 2009).

7.3.2 Technical Standards/Technical Barriers to Trade

Technical standards describe in detail the physical and chemical properties of fuels. For liquid biofuels, these have been established several years ago, whereas for solid biomass, these are often still under development.

7.3.2.1 Liquid Biofuels

Regulations pertaining to the technical characteristics of liquid transport fuels (including biofuels) exist in all countries. These have been established in large part to ensure the safety of the fuels and to protect consumers, e.g. drivers from buying fuels that could damage their vehicles' engines or power plants and pellet stove owners to buy pellets that will not damage their equipment.

Regarding the liquid biofuels, two types of technical regulations affect trade in biofuels: maximum percentages of bioethanol or biodiesel which can be mixed with petroleum fuels in the blends commercially available; and regulations pertaining to the technical characteristics of the biofuels themselves. For the former, this is a general market barrier for biofuels (e.g. the max. E15 blend wall in the US).[1] The latter type can be a technical trade barrier, i.e. technical specifications can in principle be misused to favor one (domestic) kind of fuel over another (imported) fuel.

For ethanol, although it is a single chemical compound (independently from which feedstock it is produced), there is the issue of percentage of water that used for the final product. That would differentiate Hydrous (wet) ethanol from anhydrous (dried) ethanol. In most countries in Europe hydrous ethanol is not allowed to

[1] This market barrier could be alleviated through the introduction of flexi-fuel cars, which can drive on any mix of gasoline and (hydrous) ethanol, as is commonplace in Brazil. However, this would require major changes in the car fleet and infrastructure of many countries. On April 2, 2012, EPA approved the first applications for registering ethanol for use in making E15.

be blended into gasoline. But in The Netherlands, an NTA (Netherlands Technical Agreement) has been published in Nov 2012 (NTA 8115) which allows hydrous to be blended into gasoline and can serve as a reference for changes in the European specifications (EN228). According to Dutch legislation such blending is allowed as long as final product is not called gasoline.

Biodiesel is derived from several types of feedstocks that cause variations in the chemical composition of the biodiesel (e.g. different chain lengths, varying number of double bonds), which influences its performance characteristics. The tripartite report suggests that many differences can be dealt with by blending various types of biodiesel to create an end-product that meets regional specifications for fuel quality and emissions. Other sources (Euractiv 2009; Jank et al. 2007; Oosterveer and Mol 2010) have argued that by fixing, for example, maximum levels of iodine for vegetable oils used in the EU biodiesel standard (EN 14214), the EU is limiting the use of biodiesel produced soy oil (and to a lesser extent palm oil) and is favoring rapeseed, the main European biodiesel feedstock. Similarly, the EU standard excludes lauric oil due to its cold filter plugging point (CFPP). Lauric oil is obtained during the production of biodiesel from amongst others palm kernels, effectively banning the use of palm kernel oil for biodiesel.

7.3.2.2 Solid Biomass

For wood pellets, for the EU, the CEN/TC 335 working group developed biomass standards to describe all forms of solid biofuels within Europe, including wood chips, wood pellets and briquettes, logs, sawdust and straw bales. Specifically for wood pellets, the CEN/TS 14961 standard divides wood pellets in various classes regarding size, ash content, mechanical durability etc. Next to this, the use of the ENplus certification system for residential wood pellets has increased strongly during the past years within the EU, and has largely taken over the role of many national systems such as the AustrianÖnorm M7137, the German DIN51731, and the Italian Pellet Gold systems. For industrial wood pellets, probably the most important ongoing initiative is the IWPB initiative (see below).

Different from liquid biofuels, there are no technical barriers to trade known to the authors (although sanitary and phytosanitary measures do play an important role as a barrier, see Sect. 7.3.3). However, the lack of widely-accepted technical standards for solid biomass has hampered the development of international trade in the past.

Traders buying and selling energy commodities such as oil or coal are working via screens. For these commodities, widely-accepted international technical standards and standardized contracts for buying and selling them exists, which greatly facilitates trade. So less paperwork for the same number of trades will reduce cost and increase flexibility, and may also allow for the creation of price indices. Transparent prices and products gives more comfort up- as well as down-stream investors, encouraging investments and hence growth of the market. Less risks means in practice lower prices. Transparent prices and products also provides certainty to various other stakeholders. This is extremely important, as the renewable energy business depends on the trust of a wide range of stakeholders. Ultimately, the tax payer or energy consumer is the customer, often de facto represented by a regulator and NGO's.

For all these reasons, a large number of industrial parties convened in 2010 to develop the standards for solid biomass. The steering committee of the Initiative Wood Pellet Buyers (IWPB) consisted (as of early 2013) of the seven largest wood pellets consumers in the EU (Dong Energy, Drax Power, GDFSuez, Eon, Eggborough, RWE and Vattenfall), and is supported by technical partners (SGS, CU, Inspectorate) and Argus as key publisher. Over the past years, this group has worked to develop standards regarding the following three elements:

1. the legal context/framework in different countries which a standard would have to meet;
2. the technical specifications required by the end-users, including how to measure them;
3. safeguarding the sustainability production of biomass (and, linked to the first point, meeting legal sustainability requirements)

Establishing standards is a very complex subject. Other energy carriers or also needed several years or even decades before they became a commodity, and it that sense, the standardisation of wood pellets is developing very rapidly. Participants acknowledge the relevance of transforming of the wood pellets market into a global commodity (harmonization and standardization, contractual and financial measures to increase market liquidity and price stability, etc.), as the ultimate objective, and the commitment of the players involved is very high. Nevertheless, Consensus is necessary to start up a daily trades. It is a new "territory", so it is important to think twice on various subjects and all arguments has to be considered carefully.

As of early 2013, the group has reached agreement on technical specifications; agreement on sampling standards; and an EFET (European Federation of Energy Traders)-contract is in place. Thus, significant progress has already been achieved regarding the second element. The first and third element are ongoing, as also legislation and sustainability requirements in different EU countries and under the European Commission are still under development.

7.3.3 *Sanitary and Phytosanitary Measures*

Next to technical and tariff trade barriers, a third important category are sanitary and phytosanitary measures – at least for solid, unrefined biomass. Where liquid transport fuels and refined solid biomass (such as pellets or briquettes) have undergone heat and/or high pressure treatment, unrefined solid biomass may still carry pests or pathogens which should not be propagated through international trade.

Especially wood chips are subject to such limitations: In 1984, pine wood chips from the US and Canada were found to be infested with pine wood nematodes which are considered to cause pine wilt disease. In 1985, the EU Plant Protection

Organization (EPPO) recommended that Europe as a whole bans softwood products except kiln-dried lumber from countries known to have pine wood nematodes (Bursaphelenchus xylophilus). Although the US government vehemently objected the ban, resulting in fact finding teams from Finland, Norway and Sweden visiting the US, it was implemented across the EU. This effectively prevents the trade of softwood wood chips, as the drying process would result in the burning of the wood chips. Under the EU Directive 2000/29/EC, the importation of wood chips for whatever purpose is regulated under the lumber standard of kiln drying which requires heat exposure of 56 °C for 30 min. There are several inconsistencies related to the directive: the international recognized (IPPO) standard of ISPM-15 allows for <6 mm packing material (soft or hardwood) to be shipped without further regulation, and also wood pellets from pine wood can be imported, even though technically they do not meet the heat treatment required by the kiln dried standard. However, on the short term, no changes in the directive are to be expected. wood pellets made from softwood, are literally exempt from any regulation and require no supportive scientific data for export to the EU. Therefore, there have been recent US developments on eradicating nematodes in wood chips which are less costly than kiln drying or fumigating, but have so far not led to large-scale exports of pine wood chips for energy. For more details on this topic, see also Chap. 3.

7.3.4 Health and Safety Issues When Transporting Solid Biomass Over Long Distances[2]

While the risks of shipping liquid fuels such as oil, gasoline or LNG have been known and dealt with for decades, transporting relatively new solid biomass types in large amounts has led to the identification of specific health and safety issues. This section gives an overview of some of the main occupational hazards in transportation, handling and storing of solid biomass. The size, shape, moisture content and the type of raw material directly influence the transport, handling and storage properties of a solid biofuel. Types of solid biofuels can be specified in accordance with ISO 17225 based on traded forms.

The properties of a biomass material and the intended use determine how the material should be safely produced, transported, stored and used. While woody biofuels such as pellets and chips from fresh or recycled wood dominate the market in terms of volumes, other solid biofuels such as straw, biodegradable fuels used for anaerobic digestion and municipal solid waste pose specific health and safety challenges that need to be addressed.

Self-heating processes may be due to biological metabolic reactions (microbiological growth), exothermic chemical reactions (chemical oxidation) and heat-producing

[2]This section is an extract from the following publication: *Health and Safety Aspects of Solid Biomass Storage, Transportation and Feeding. Produced by IEA Bioenergy Task 32, 36, 37 and 40. February 2013. Edited by Jaap Koppejan, Procede Biomass BV, The Netherlands.*

physical processes (e.g. moisture absorption), and it may occur both for dry and wet biofuels. It may become problematic if a pile or silo is so large that the heat generated cannot be easily dissipated to the surroundings. While this is not the case for relatively small scale installations as e.g. used by households, it needs attention for larger industrial storages. Several test methods are available for determining self-heating potential and self-ignition of materials on small scale, which can then be extrapolated to predict self-heating potential at larger scale. Apart from self-heating, biomass stock may be set on fire through various external sources such as hot bearings, overheated electric motors, back-fire, etc.

Off-gassing is the process where volatile organic compounds are released in the logistical chain. One mechanism is the initial release of lipophilic compounds, yielding carbonyl compounds (aldehydes and ketones) and also complex terpenes. CO, CO_2 and CH_4 may also be released. The concentrations of aldehydes found in domestic sites, warehouses and ships constitute a health hazard and require attention and preventive measures to be taken. Hexanal may enter the body by contact with skin or by inhalation and cause skin irritation, headaches, and discomfort on eyes and nose. Other aldehydes such as methanal and ethanal are suspected to be carcinogenic in high doses and may also have some short time effect on human health. There are several guidelines issued by government official institutes that describe the effect these aldehydes have on human health depending on exposure time and level. Monoterpenes (particularly present in fresh raw material) cause eyes and respiratory system irritation. CO may be released from the auto oxidation of lipophilic compounds. Related hazards are predominantly poisoning, but it may also contribute to self-heating or ignition processes. A combination of proper ventilation, gas meters and the use of self-contained breathing devices is needed in areas where the levels of CO might increase to poisonous concentrations.

Oxidation of fatty acids in sawdust and other moist fuels is accelerated by microbial activity with mesophilic bacteria and fungi up to approximately 40 °C and by thermophilic bacteria up to approximately 70 °C. Above this temperature chemical oxidation becomes dominant and further raises the temperature, in many cases up to an uncontrolled temperature range.

Dust clouds are a major cause of damage in the bioenergy sector. The combination of relatively small particle sizes and low minimum ignition energy results in a high ignition sensitivity. Significant amounts of factory dust may stay be suspended in the air, so that the Minimum Explosible Concentration is easily reached under practical conditions if cleaning and ventilation are insufficiently done. It is therefore important to minimize the risk of dust explosions, by minimizing the risk of sparks (a.o. due to electrostatic discharge through proper grounding) and good dust housekeeping through dust prevention and dust collection. Once an explosion takes place, it needs to be properly contained, suppressed or vented. Compliance with ATEX Directives and NFPA guidelines is essential in this respect.

The health risks poses by biomass fuels in the form of dusts and bioaerosols come from the both the physical particle and size effects. As particles become smaller they pose greater hazard. As a result limits on PM10 and PM2.5 (particles

Table 7.1 Results from measurements in five transatlantic ships with wood pellets

Ship no.	1	2	3	4	4	5	5
Place	Cargo hold	Cargo hold	Cargo hold	Stairway	Stairway	Cargo hold	Stairway
Date	Nov 2006	Jan 2007	Feb 2007	Feb 2007	March 2007	Oct 20007	Oct 2007
CO ppm	5,850	6,980	14,650	10,960	11,510	11,950	7,710
CO_2 ppm	9,340	3,240	7,070	5,450	5,160	21,570	12,360
O_2 %	n/a	n/a	n/a	n/a	5	0,8	8,4

Svedberg et al. (2008) and Svedberg, Sundsvalls Hospital, personal communication, 2012

less than 10 or 2.5 μm respectively) are becoming more prevalent in national regulations. In addition the organic nature of biomass fuels may result in additional impacts through either allergenic or pathogenic routes. The most prevalent feature will be the allergenic responses and the majority of the effects will be minor and short lived; but increasing severity of impact will also be linked to falling incidence of response. In the same way pathogenic responses will be a rare occurrence, but potentially result in severe hazards.

Current a large proportion of the wood pellets produced worldwide are shipped by ocean vessels. The safety is regulated by the International Maritime Organization (IMO) and there are significant safety issues onboard vessels, in terminal storage and during handling. Several fatal accidents have been recorded, which have resulted from exposure to harmful gas emissions from biofuels, particularly in enclosed spaces such as storage silos, flat storage buildings and cargo holds in ocean vessels. Table 7.1 provides measurements of toxic gases and oxygen content in cargo holds in vessels during Atlantic crossing.

The IMO regulate the transportation of pellets in ocean vessels and prescribe conditions under which wood pellets can be carried. Cargo holds are sealed during ocean voyage which results in very fast oxygen depletion and generation of CO, CO_2, CH_4 and some H_2. Entry in to cargo holds and communicating spaces are prohibited unless the spaces have been thoroughly ventilated and the gas concentration has been verified by a combination of oxygen and CO measurements (Svedberg et al. 2008; Svedberg 2012; Svedberg, Sundsvalls Hospital, personal communication, 2012; Christensen 2012; Dahl 2012).

7.3.5 *Possible Impact of Sustainability Criteria on Trade*

Sustainability criteria and certification schemes are discussed in more detail in Chap. 6, but should again briefly mentioned here, as they can under certain circumstances affect bioenergy trade in several different ways:

- Elaborate and comprehensive sustainability criteria and corresponding certification schemes might exclude small-scale farmers because they are dominantly designed for large-scale agro-industry. Many certification schemes are data- or information-intensive and require costs and capacities that are often out of reach for most smallholders. As a recent FAO report (Elbehri et al. 2013) states "the schemes, to the extent that they are established to control imports, can hinder trade and reduce market access – especially for developing countries with comparative advantages in business production, and which see in this industry a real opportunity for development and for overcoming rural poverty and high unemployment … Many developing countries express concern that certification schemes can become indirect trade barriers when not managed properly".

- Also, while the introduction of e.g. the mandatory sustainability criteria under the RED has led to some harmonisation for liquid biofuel certification schemes, there are still many different systems around with differing criteria and indicators. For solid biomass, the same problem exists, but may be at least to some extent be solved through the inclusion of sustainability criteria in the Enplus and IWPB certification systems (see previous section).
- Changing sustainability criteria also have a profound impact on the industry. For example, with the establishment of sustainability criteria in the RED for liquid biofuels, many biofuel producers deemed it certain that compliance with these criteria would guarantee long-term market access. However, with the recent debate regarding indirect land use change (iLUC) and the ensuing proposal of the EC to limit food crops to meet only half of the total 10 % renewable fuel target has caused significant concern amongst the industry. Similarly, the on-going scientific insights an discussions regarding e.g. the definition of 'primary forests' (IEA Bioenergy 2013) and perhaps even more significant correct carbon accounting of forest biomass (see e.g. Lamers and Junginger 2013) have increased uncertainty amongst industrial stakeholders, discourage new investments in solid biomass conversion capacity, and ultimately may act as indirect barriers for solid biomass trade. As of mid-2013, no mandatory sustainability criteria are likely to be implemented on the short term on an EU level, and it is the authors' expectation that countries such as Germany, Belgium, the UK and the Netherlands will investigate a cooperation to introduce regulation regarding national mandatory criteria.
- On the other hand, legislation and mandates with binding sustainability criteria may also increase trade flows. One example is the import of lignocellulosic ethanol from Brazil to the US, as it is considered as "non-cellulosic advanced biofuels" (with at least 50 % GHG reduction). With an requirements for this category expected to reach 13.2 billion litres in 2020, the Brazilian ethanol might become the main supplier within this category (see also Sect. 7.2.1 of this chapter and Chap. 4).

7.4 Discussion and Outlook

As shown in this chapter, there are several drivers responsible for the strong increase in biomass trade over the past decade. Most of them are likely to remain: concerns regarding the effects of climate change remain unchanged, and policy targets for renewable energy for 2020 have so far remained (largely) intact despite the economic crisis. At the same time, the list of barriers potentially hampering the further growth is long and very heterogeneous; varying from classical tariff barriers and technical trade barriers (mainly liquids) to the lack of standardized contracts and mandatory sustainability requirements, health and safety issues to overcome and phytosanitary requirements (mainly solid biomass).

The global sustainable bioenergy sector is complex and heavily dependent on political developments. Global bioenergy trade of solids and liquids may continue to grow, as the economic drivers will influence primary or secondary capital efficiency of ventures and projects.

For bioenergy trade towards the EU to grow further, long-term investment security, including a support regime for the unprofitable part of energy production from biomass compared to its fossil competitors is required. That can be in the form of subsidies, but also in the form of a producer-and/or suppliers obligation. However, such a system would ideally have to be implemented on an European scale – national taxes will not bring any solution. Similarly, a well-functioning ETS on an EU-level is an important precondition. Also, uniform, mandatory and sustainability criteria in the EU for solid and liquid biofuels, but also guaranteed not to change, at least until 2020. Last but not least (and perhaps most importantly), consumers must be willing to pay extra for renewable energy.

The main expected barriers for the coming years are that the solid biomass market will not become a commodity is in the short term, i.e. that the legal and technical frameworks are not sufficiently standardized. As a consequence, trade may remain limited and risks may be insufficiently covered. Furthermore, there is a real risk that no agreement can be reached amongst the EU (and international) stakeholders on how to deal with sustainability. On the other hand, as shown with the mid-2013 decision to remove significant amounts of emission rights from the ETS, other multi-year (EU)/national governmental policies may also be successful.

Nevertheless, there is a wide-spread understanding and recognition of the importance of renewable energy sources (including biomass) by many countries and dominant global players, and a belief that this fast growing business could not be reversed. The key reason is not only the mitigation of climate change but also for reasons of energy security, economic growth and employment.

Due to the crisis, since 2008, many unforeseen effects and chain reaction effects occur. Despite the peak in 2008 and the volatility on the commodity markets, expectations are that the markets will level out on a lower level than before the peak. Looking at the sensitivity analysis and the needed drivers mentioned above, bioenergy projects can become profitable on the mid-term. On the other hand, production facilities are currently decreasing their production volumes due to lower product prices. Consequently, the CO_2 emissions are decreasing and thus their price is declining. Coal demand is decreasing (also due to US shale gas, another unexpected development), and thus the tension on this market is getting less. Shipping prices are declining. The spread between coal plus CO_2 and solid biofuels is getting smaller, so the attractiveness of sustainable projects is getting lower, also based on the currently relatively unstable regulatory framework. The crisis could therefore influence negatively the climate change business in the short term – but will profit a lot on the mid and long term. From a business point of view, now may also be a time to get in low and profit on the mid-term from increasing business opportunities and 'to ride the wave'. Under a stable regulatory framework, even if in short term profit is slow, companies with a long term vision would still find sustainable projects attractive enough to invest.

References

AFDC. (2013). *Renewable fuel standard*. Alternative fuels data center. US Department of Energy. Available at: http://www.afdc.energy.gov/laws/RFS

Biomass Energy Centre. (2013). *Co-firing*. Available at: http://www.biomassenergycentre.org.uk/portal/page?_pageid=75,41175&_dad=portal&_schema=PORTAL

Christensen, M. S. R. (2012). *Personal communication. Storage logistics and sales manager, Copenhagen Merchants*, Charlottenlund, Denmark.

Dahl, J. (2012). *Project: Large scale utilization of biomass pellets for energy applications*. Acronym: LUBA. Ongoing project. Danish Technological Institute.

Dufey, A. (2007). *International trade in biofuels: Good for development? And good for environment?* International Institute for Environment and Development. Available at: www.iied.org

EIA. (2013). *Spot prices for crude oil and petroleum products*. Available at: http://tonto.eia.doe.gov/dnav/pet/xls/pet_pri_spt_s1_m.xls

Elbehri, A., Segerstedt, A., & Liu, P. (2013). *Biofuels and the sustainability challenge. A global assessment of sustainability issues, trends and policies for biofuels and related feedstocks*. Rome: FAO. Available at: http://www.fao.org/docrep/017/i3126e/i3126e.pdf

EPA. (2013). *Renewable Fuel Standard (RFS)*. Available at: http://www.epa.gov/otaq/fuels/renewablefuels/index.htm

Euractiv. (2009). *Dossier biofuels, trade and sustainability*. Last update 29 July 2009. Available at: http://www.euractiv.com/en/trade/biofuels-trade-sustainability/article-171834S

EurObserv'ER. (2009). Biofuelsbarometer. *SystemesSolaires, le Journal des energies renouvelables, 192*. Available at: http://www.energies-renouvelables.org

Farm Futures. (2013). *Ethanol groups intend to challenge EU tariff*. Available at: http://farmfutures.com/story-ethanol-groups-intend-challenge-eu-tariff-0-95132

Heinimö, J., & Junginger, M. (2009). Production and trading of biomass for energy—An overview of the global status. *Biomass and Bioenergy, 33*(9), 1310–1320.

Hoefnagels, R., Junginger, M., Resch, G., Matzenberger, J., Panzer, C., & Pelkmans, L. (2011). *Development of a tool to model European biomass trade* (Report for IEA bioenergy task 40). Available at: http://www.bioenergytrade.org/downloads/development-of-a-tool-to-model-european-biomas.pdf

IEA Bioenergy. (2013). *The science-policy interface on the environmental sustainability of forest bioenergy*. A strategic discussion paper. IEA Bioenergy: ExCo:2013:03. Available at: http://www.ieabioenergy.com

Indexmundi. (2013). Available at: http://indexmundi.com. Accessed 29 Apr 2013.

Irwin, S., & Good, D. (2012, December 7). What's driving the surge in ethanol imports? *FarmDoc Daily*, Department of Agriculture and Consumer economics, University of Illinois Urbana-Champaign. Available at: http://farmdocdaily.illinois.edu/2012/12/whats-driving-the-surge-in-ethanol.html

IWPB. (2013). *Initiative wood pellet buyers. Standardizing industrial wood pellet trading in Europe*. Available at: http://www.laborelec.be/ENG/initiative-wood-pellet-buyers-iwpb/

Jank, M. J., Kutas, G., do Amaral, L. F., & Nassar, A. M. (2007). *EU and U.S. policies on biofuels: Potential impacts on developing countries*. Washington, DC: German Marshall Fund of the United States. Available at: https://ees.ucsb.edu/academics/documents/EU_US_biofuels_policies_Marshall_Fund.pdf

Junginger, M., van Dam, J., Alakangas, E., Virkunnen, M., Vesterinen, P., & Veijonen, K. (2010, February). *Solutions to overcome barriers in bioenergy markets in Europe- D2.2. Resources, use and market analysis*. EUBIONETIII – Solutions for biomass fuel market barriers and raw material availability – IEE/07/777/SI2.499477. VTT/Utrecht University.

Junginger, M., van Dam, J., Zarrilli, S., Ali Mohamed, F., Marchal, D., & Faaij, A. (2011). Opportunities and barriers for international bioenergy trade. *Energy Policy, 39*, 2028–2042.

Kfouri, G. (2012, March 13). *EU implements higher import taxes for ethanol fuel blends with 30% gasoline*. Platts. Available at: http://www.platts.com/RSSFeedDetailedNews/RSSFeed/Petrochemicals/8050604

Koplow, D. A. (2009). *Boon to bad biofuels*. Washington, DC: Earth Track and Friends of the Earth. Available at: http://www.foe.org/news/archives/2009-05-report-a-boon-to-bad-biofuels-released

Lamers, P., & Junginger, M. (2013). The 'debt' is in the detail: A synthesis of recent temporal forest carbon analyses on woody biomass for energy. *Biofuels, Bioproducts and Biorefining, 7*(4), 373–385. doi: 10.1002/bbb.1407

Londo, M., Lensink, S., Wakker, A., Fischer, G., Prieler, S., van Velthuizen, H., de Wit, M., Faaij, A., Junginger, M., Berndes, G., Hansson, J., Egeskog, A., Duer, H., Lundbaek, J., Wisniewski, G., Kupczyk, A., & Könighofer, K. (2010). The REFUEL EU road map for biofuels in transport: Application of the project's tools to some short term policy issues. *Biomass and Bioenergy, 34*(2), 244–250. doi:10.1016/j.biombioe.2009.07.005.

Murphy, S. (2008, April). *The multilateral trade and investment context for biofuels: Issues and challenges*. Institute for Agriculture and Trade Policy (IATP). Available at: http://www.iatp.org/documents/the-multilateral-trade-and-investment-context-for-biofuels-issues-and-challenges

Oosterveer, P., & Mol, A. P. J. (2010). Biofuels, trade and sustainability: A review of perspectives for developing countries. *Biofuels, Bioproducts and Biorefining, 4*(1), 66–76.

Steenblik, R. (2007, December). *Subsidies: The distorted economics of biofuels* (Discussion Paper No. 2007-3). Geneva: The Global Subsidies Initiative (GSI), International Institute for Sustainable development (IISD).

Svedberg, U. (2012). Faror och hälsorisker vid pelletslagring (Dangers and risks at storage of pellets). In *Pellets 2012*. Stockholm.

Svedberg, U., Samuelsson, J., & Melin, S. (2008). Hazardous off-gassing of carbon monoxide and oxygen depletion during ocean transportation of wood pellets. *Annals of Occupational Hygiene, 52*(4), 259–266.

Zakaria, A., Wakker, E., & Theile, C. (2009). *Failing governance—avoiding responsibilities European biofuel policies and oil palm plantation expansion in Ketapang District, West Kalimantan (Indonesia)*. Friends of the Earth Netherlands (Milieudefensie) and WALHI Kalimantan Barat, Indonesia. Available at: http://www.foei.org/en/resources/publications/pdfs-members/agrofuels/european-biofuel-policies-failing-governance-avoiding-responsibilities

Zarrilli. (2008). *Making certification work for sustainable development: The case of biofuels* (UNCTAD/DITC//TED/2008/1).

Chapter 8
Medium and Long-Term Perspectives of International Bioenergy Trade

Lukas Kranzl, Vassilis Daioglou, Andre Faaij, Martin Junginger, Kimon Keramidas, Julian Matzenberger, and Erik Tromborg

Abstract In the coming decades, huge challenges in the global energy system are expected. Scenarios indicate that bioenergy will play a substantial role in this process. However, up to now there is very limited insight regarding the implication this may have on bioenergy trade in the long term. The objectives of this chapter are: (1) to assess how bioenergy trade is included in different energy sector models and (2) to discuss the implications and perspectives of bioenergy trade in different energy scenarios. We grouped scenarios from the models IMAGE/TIMER, POLES and GFPM according to their policy targets and increase of bioenergy use in "ambitious" and "moderate" bioenergy scenarios and compared results regarding bioenergy trade for solid and liquid biomass. Trade balances for various world regions vary significantly in the different models and scenarios. Nevertheless, a few robust trends and results can be derived up to the year 2050: Russia and former USSR countries could turn into strong biomass exporting countries. Moreover, Canada, South-America, Central and Rest-Africa as well as Oceania could cover another substantial part of the bioenergy supply. As importing countries, India, Western Europe and China might play a key role. The results show (1) the high relevance of the topic, (2) the high uncertainties, (3) the need to better integrate social, ecological, economic

L. Kranzl (✉) • J. Matzenberger
Institute of Energy Systems and Electrical Drives,
Vienna University of Technology, Vienna, Austria
e-mail: kranzl@eeg.tuwien.ac.at

V. Daioglou • A. Faaij • M. Junginger
Copernicus Institute, Utrecht University, Utrecht, The Netherlands

K. Keramidas
Enerdata, Paris, France

E. Tromborg
Department of Ecology and Natural Resource Management,
Norwegian University of Life Sciences, Ås, Norway

M. Junginger et al. (eds.), *International Bioenergy Trade: History, status & outlook on securing sustainable bioenergy supply, demand and markets*, Lecture Notes in Energy 17,
DOI 10.1007/978-94-007-6982-3_8,

and logistical barriers and restrictions into the models and (4) the need to better understand the potential role of bioenergy trade for a sustainable, low-carbon future energy system.

8.1 Global Scenarios of Bioenergy Demand and the Question of Bioenergy Trade

In the coming decades big changes in the global energy system are expected. On the one hand, the insecure supply of fossil fuels might have a major impact. On the other hand, if the global targets of climate change mitigation are taken seriously, a huge transition of the overall energy system will be required. Therefore, a substantial effort has been taken in the last years to improve the modelling of regional and global energy systems, develop scenarios and show options and impact of climate mitigation measures. Bioenergy plays a crucial role in these studies and scenarios.

The IPCC report on renewable energy (IPCC 2011) includes a comprehensive comparative analysis of global energy scenarios. In particular, this investigation includes an assessment of the relevance of biomass in 137 scenarios up to 2050. Starting from the historic level of 50 EJ (about 1,200 Mtoe) of bioenergy use in 2008, most of the scenarios show a considerable increase of bioenergy use. The median of mitigation scenarios (440–600 ppm and <440 ppm) show an increase to about 70–80 and 120–155 EJ in 2030 and 2050, respectively.

According to the IEA World Energy Outlook 2012, primary demand for modern bioenergy in the scenario "new policy" will more than double up to the year 2035. Moreover, the patterns of bioenergy use are expected to change substantially. Power generation and production of biofuels for transport will constitute a larger share of biomass use.

Such a strong expansion of bioenergy use in the next decades requires the exploitation of additional bioenergy sources. A significant part of these potential sources might be not located in the same region or even continent where the demand takes place. So, it seems obvious that these scenarios would imply an impact on the regional balance of demand and supply of bioenergy leading to a change in trade patterns in various world regions. However, up to now these implications on regional supply and demand gaps and the related trade of bioenergy have been documented only in a very few cases (e.g. WEO 2012; Raunikar et al. 2010).

Only recently, with the new World Energy Outlook 2012, a bioenergy trade module has been implemented in the IEA World Energy Model (WEM). The results show that the volumes, routes, fuels, logistics of bioenergy trade will hence look quite different to what we are used to today. However, different scenarios and models of the global bioenergy sector do not show the same picture of the future development of bioenergy use. The perspectives of bioenergy trade depend on different scenarios and impacts regarding energy markets, technological development, energy and climate change policies.

Despite of the crucial role of bioenergy for future energy scenarios, bioenergy trade between countries and world regions usually is not investigated and documented explicitly. Compared to the profound global energy modelling approaches, the analyses of future perspectives of bioenergy trade are still in an early phase of their development. In this chapter, we take a closer look at the implications of global energy model scenarios on bioenergy trade. Based on a broad set of scenarios, we discuss perspectives of international bioenergy trade in the coming decades, we derive robust trends and corresponding conclusions.

The main objective of this chapter is twofold:

1. to assess how bioenergy trade is included in different energy sector models covering bioenergy, and
2. to analyse the implications and perspectives of bioenergy trade in different energy market scenarios

The comparative investigation of models and scenarios of international bioenergy trade leads to a higher insight of patterns of bioenergy trade, drivers and dependencies. So, the objective is to learn about the key linkages, relations and interdependencies. The quantification of traded volumes in various scenarios is an important aspect, but not the primary objective of this chapter. Moreover, we want to emphasize that our objective here is not a comprehensive model comparison of selected energy models. This would require much higher effort in terms of defining consistent framework conditions and investigate model results in terms of the specific modeling approach.

We will start with a brief overview of the reviewed studies. A few of these studies have been selected for a closer investigation. For these scenarios, we provide a comparison of scenarios which leads to an analysis of model based drivers of bioenergy trade.

The work presented here is based on a literature review of articles and reports that describe quantitative models with an international coverage that include bioenergy. The more in depth analysis of selected models is based on data made available by researchers involved in the given models.

8.2 Global Models of Bioenergy Trade

Many studies have been undertaken to assess the potential of biomass to contribute to future energy supply. A smaller number of studies deals with the gap between regional bioenergy demand and supply and bioenergy trade. Conclusions from these studies vary significantly. We have indentified 28 models and/or studies dealing with bioenergy trade in some form. The models have different scopes with some focusing specifically on biomass markets and trade, while others are more general energy models which also include international bioenergy trade. For this chapter, we screened these models and studies in order to identify models suited for in-depth analysis of international biomass trade. Out of the 28 identified bioenergy trade models, 22 models were selected according to their potential to model global bioenergy trade on a sufficient regional resolution (Fig. 8.1). Relevant literature which has been screened for this task was: (Berndes et al. 2003; Bouwman et al. 2006; Gielen et al. 2003; Hamelinck and Hoogwijk 2007; Hansson and Berndes 2009;

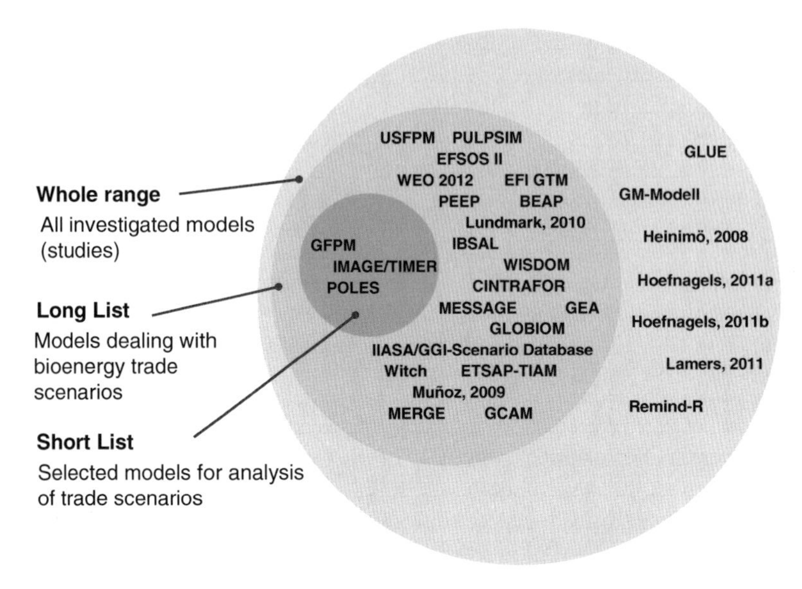

Fig. 8.1 Three-stage model selection process

Havlík et al. 2011; Heinimö et al. 2008; Hoefnagels et al. 2011a, b; Ince et al. 2011; Kallio et al. 2004; Lamers et al. 2011; Lundmark 2010; Masera et al. 2006; Muñoz et al. 2009; Smeets et al. 2007; Sokhansanj et al. 2006; Szabó et al. 2009; Yamamoto et al. 2001).

The models identified for further analysis ("long-list") have been characterized according to specific criteria regarding bioenergy trade, based on available literature and, where not specified in sufficient detail, a questionnaire has been sent out to the respective modelling groups. The following criteria for model selection have been analysed: the extent to which the model does cover biomass trade and if regional or global trade patterns are assumed, sectoral coverage, geographical regional aggregation and scenario time frame. In most models a scenario timeframe until year 2100 is considered. More detailed information on these models is provided in Matzenberger et al. (2013).

8.2.1 Selected Models for Scenario Comparison

Three models have been selected for a detailed comparison of scenarios and their impact on global bioenergy trade: GFPM, TIMER and POLES.

The TIMER model is a dynamic energy system simulation model developed by the Netherlands Environmental Assessment Agency (PBL) (van Vuuren et al. 2007). It has bottom-up engineering information as well as top-down investment behavior rules and technological change. It is part of the larger integrated assessment model IMAGE, from which it gets its biophysical data and in turn provides energy and industry related emissions. The simulation process is dynamic recursive on a year-by-year basis. Energy demand is determined from economic activity and population increase and is calculated over five end-use sectors (Industry, transport, residential, services and 'other') as well as three transformation sectors (electricity, hydrogen and heat). This energy demand can be met from a number of energy carriers which compete with each other on a relative cost basis. Thus demand for energy carriers including bioenergy is price elastic. Primary biomass resources include energy crops (sugar, starch, woody) as well as organic residues.

The POLES model provides a complete system for the simulation and economic analysis of the sectoral impacts of climate change mitigation strategies. The POLES model is not a General Equilibrium Model, but a dynamic Partial Equilibrium Model, essentially designed for the energy sector but also including other GHG emitting activities, with the 6 GHG of the "Kyoto basket". The simulation process is dynamic, in a year by year recursive approach of energy demand and supply, with lagged adjustments to prices and a feedback loop through international energy prices that allow describing full development pathways from 2005 to 2100. There is an explicit breakdown of total land surface across main categories for each country/region of the model. Primary biomass resources have been divided into three categories (forest residues, short rotation crops, other energy crops like sugar or bio-oil crops).

The GFPM model is a spatial partial equilibrium model for forest products based on price endogenous linear programming (Buongiorno et al. 2003). In the analyses of bioenergy development, fuelwood demand in each country is represented by a price-elastic demand function with exogenously specified long-run shifts of demand based on scenario assumptions about global expansion in fuelwood consumption (Raunikar et al. 2010; Buongiorno et al. 2011). Demand for final products are defined by demand at last year's price and price elasticity of demand. Demand changes in each country due to changes in GDP and elasticity of demand with respect to GDP. The model is limited to the forest and forest biomass sectors and covers 14 principal categories of forest products of which fuelwood includes wood used for heating, cooking, power and fuel production.

8.2.2 Scenario Settings

POLES and TIMER scenarios correspond closely to each other. The scenarios presented for the TIMER model are based on the OECD environmental outlook (OECD 2012). In all cases, the development of population growth, GDP growth, land availability and crop yields are the same. Since in the TIMER model energy demand is elastic to energy prices, total final energy demand varies across scenarios. The results of POLES are derived from scenarios initially developed for EMF27. Population & GDP are consistent across all scenarios. The main scenarios are the following:

- *Reference scenario*: OECD Environmental Outlook baseline for TIMER and BAU for POLES.
- *Scenarios for different levels of CO_2-prices*: 20$\$_{2005}$/t$CO_2$ & 100\$/tCO_2: Global carbon tax is applied to the carbon content of fuels instantaneously in 2015 and

remains throughout simulation period. All energy consuming sectors and fuels are affected.

- *Climate mitigation scenarios:* resulting in different CO_2-concentration levels of 450 and 650 ppm, respectively. Global carbon taxes are gradually applied uniformly across all fuels and sectors in order to ensure carbon concentration targets are met.
- *Trade barriers:* Transaction costs for bilateral trade are increased to such a level that interregional trade becomes unattractive. Thus, bioenergy consumption is stronger limited by local production.

GFPM-Scenarios are based on Buongiorno et al. (2011):

- *IPCC scenario A1B/high fuelwood demand*: Continuing globalization would lead to high income growth and low population growth, and thus the highest income per capita by the year 2060. Eighty percent increase in biofuel demand up to 2030 from 2006.
- *Low fuelwood demand*: 20 % increase in fuelwood demand up to 2030, other assumptions as in the high fuelwood demand scenario (50 % of the fuelwood demand growth in scenario A1B).

More details regarding the scenario settings and assumptions are documented in Matzenberger et al. 2013.

8.3 Perspectives of International Bioenergy Trade: Selected Scenarios

The model scenarios outlined above have been compared in terms of the following results:

- Global bioenergy demand and production,
- Bioenergy demand and production in 20 world regions,
- Net trade balance of bioenergy in 20 world regions.

Key results of this comparison are shown in the following graphs. Figure 8.2 shows that scenarios lead to significant growth of bioenergy production and demand on a global scale. The current level of about 50 EJ (1.2 Gtoe) of world bioenergy production increases to a level of up to 150–170 EJ (3.6–4.1 Gtoe) in 2050 and 170–220 EJ (4.1–5.3 Gtoe) in 2070. However, it is not only the amount of bioenergy, also the structure of bioenergy use and mix of resources, fuels and conversion technologies changes. Traditional biomass reduces in all scenarios and step by step is replaced by modern biomass. The growth is clearly on solid biomass resources (the values for liquid biomass in the TIMER and POLES scenarios include second generation biofuels from solid biomass).

A few scenarios also indicate less growth in bioenergy demand, in particular, TIMER environmental outlook and other scenarios with low or moderate climate mitigation policies (20$/t CO_2 scenarios and 600 ppm concentration levels). Of course, also the regional distribution of supply and demand and thus trade balances vary among those scenarios. Thus, we distinguished moderate and ambitious

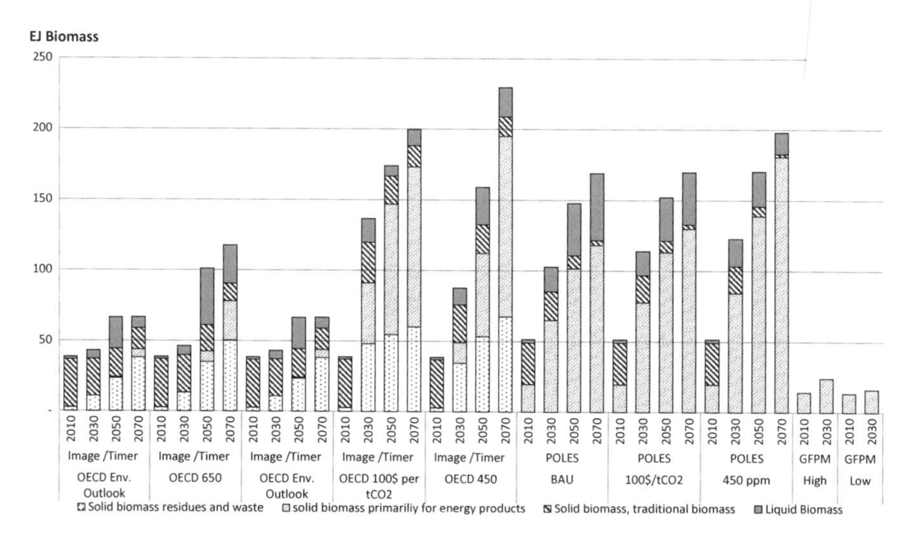

Fig. 8.2 World bioenergy production in selected scenarios. Note: GFPM covers forestry products only. Traditional biomass is only distinguished in TIMER and is not considered in GFPM

bioenergy scenarios. The ambitious scenarios comprise those achieving the 450 ppm scenario or assuming a carbon price of 100$ per t CO_2, i.e.:

- TIMER: OECD 450 ppm scenario, OECD 100$ per t CO_2 scenario
- POLES based on EMF scenarios: 450 ppm, 100$ per t CO_2
- GFPM: high

The other scenarios are grouped as "moderate" bioenergy scenarios:

- TIMER: OECD environmental outlook, OECD EO trade barriers, OECD 650 ppm, OECD 20$ per t CO_2
- POLES: based on EMF scenarios G1 Reference, G4 BAU, BAU + trade barriers, 650 ppm, 20$ per t CO_2
- GFPM: low

Global overall bioenergy demand in moderate bioenergy scenarios is distributed more evenly than in the ambitious scenarios. In the average of the moderate scenarios the regions USA, Central and Rest Africa, Western Europe, India, China and South East Asia show a demand in the range of 6–10 EJ (140–240 Mtoe) and 10–16 EJ (240–382 Mtoe) in 2030 and 2050, respectively. In contrast, the ambitious scenarios are dominated by the demand in India and China (14–17 EJ and around 25 EJ in 2030 and 2050, respectively). China also shows the largest range within the investigated scenarios: Ambitious scenarios result in a range of 20 to more than 40 EJ in 2050. This overall increase of bioenergy demand is developing in a different way for liquid and solid fuels: The share of liquid biofuels on total bioenergy (sum of solid and liquid) for the median of the ambitious scenarios is 18 % (2030) and 14 % (2050). The maps in Fig. 8.3 show the global distribution of bioenergy demand for solid and liquid biofuels in the median of ambitious scenarios.

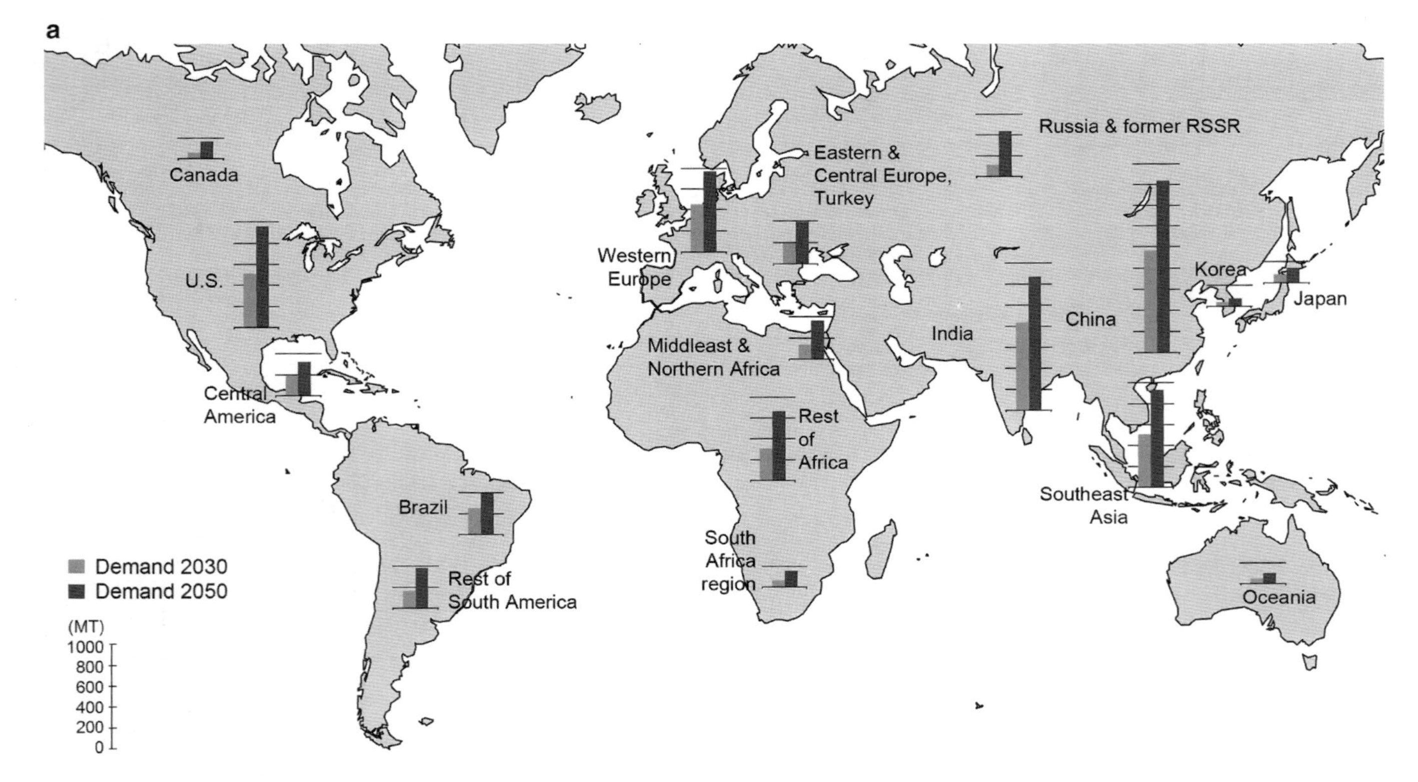

Fig. 8.3 Regional bioenergy demand in the median of ambitious model scenarios 2030 and 2050. (**a**) solid biomass, (**b**) liquid biomass (Unit: Mt)

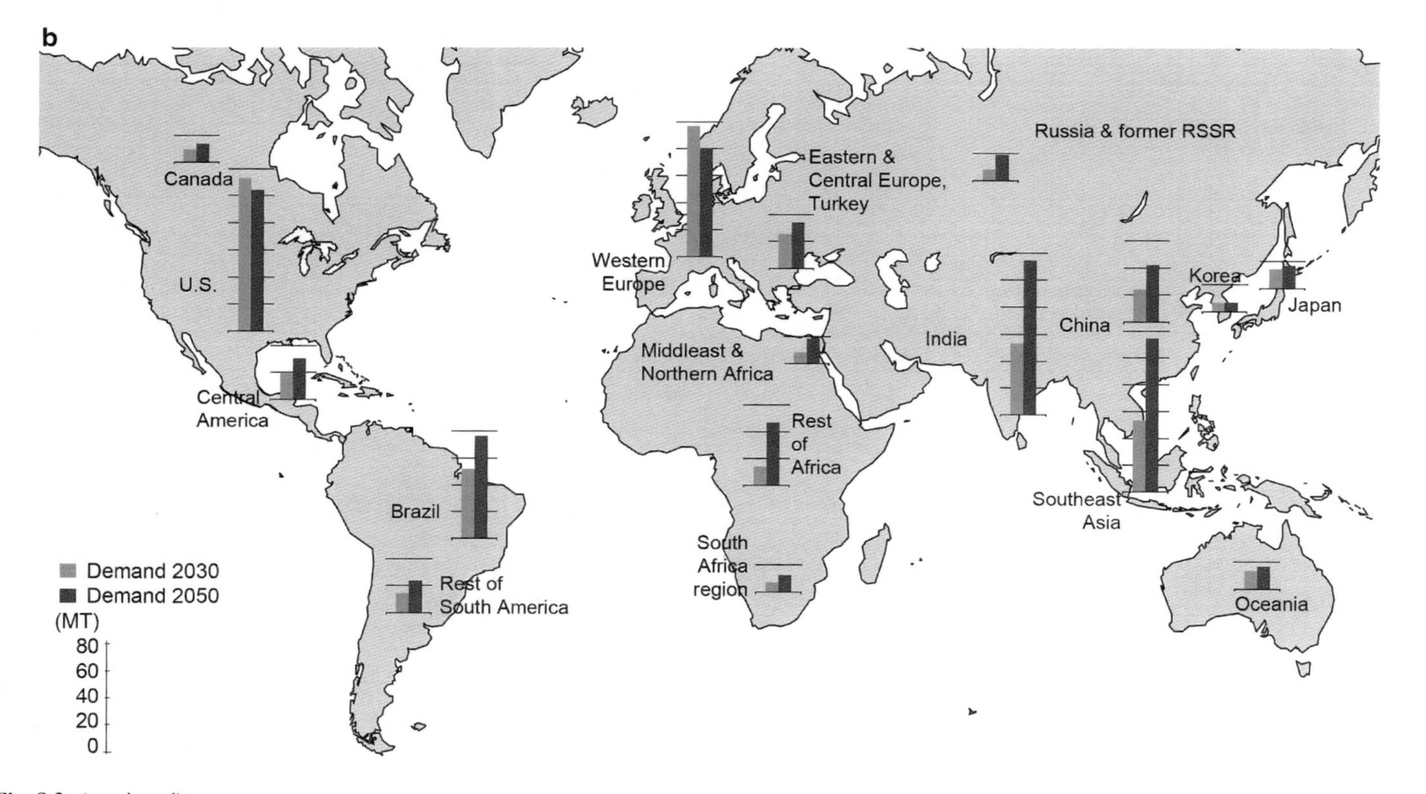

Fig. 8.3 (continued)

So, where does the biomass come from? To which extent are these regions with a high bioenergy demand depending on biomass imports? The following figures show the range of trade balances for selected ambitious and moderate scenarios (Fig. 8.4).

In ambitious scenarios, 14–26 and 14–30 % of global bioenergy demand is traded between regions in 2030 and 2050, respectively. In more detail, the model scenarios show a huge range of potential bioenergy trade: for solid biomass, in ambitious scenarios bioenergy trade ranges from 700 Mt to more than 2,500 Mt in 2030 and from 800 Mt to almost 4,200 Mt in 2050. These values only take into account TIMER and POLES scenarios since GFPM covers forest products only. In the scenario "high" of GFPM, in 2030, 25 % of forest based global bioenergy demand is traded between world regions. For liquid biomass, the ambitious scenarios show a bioenergy trade in the range of 65 Mt to more than 360 Mt in 2030 and from 40 to 520 Mt in 2050.

In moderate scenarios, 0–20 and 7–26 % of global bioenergy demand is traded between regions in 2030 and 2050, respectively. For solid biomass, this corresponds to an amount of 3–1,500 and 100–2,000 Mt in 2030 and 2050, respectively. These values only take into account TIMER and POLES scenarios since GFPM covers forest products only. In the scenario "high" of GFPM, in 2030 21 % of global bioenergy demand from forestry products is traded between world regions. For liquid biomass, the range of bioenergy trade in moderate scenarios amount to 1–360 and 12–820 Mt in 2030 and 2050, respectively.

For comparison, as shown in Chap. 2, trade volumes of liquid fuels (ethanol and biodiesel) did not exceed 5 Mt in 2011. Net woody biomass trade in 2010 amounted to roughly 18 Mt (mainly wood pellets fuel wood and wood waste). Thus, the model results show a huge increase of bioenergy trade in the coming decades in most of the scenarios (in particular in the more ambitious bioenergy scenarios).

For a proper interpretation of these results, one should take into account that these values underestimate the international trade that would actually occur for the following reasons: (1) trade streams are only reported between world regions; most of these world regions consist of a number of different countries with export and import activities between individual countries which are not estimated in this analysis; (2) only net trade balances are reported; whereas in reality, often both import and export between two regions are observed.

8.4 Synthesis and Conclusions: Future Challenges and Perspectives of Global Bioenergy Trade

8.4.1 Analysis and Discussion of Robust Trends and Trade Patterns

The models and scenarios show considerable differences for bioenergy demand and for trade balances in different world regions. Nevertheless, the results shown above allow us to derive some robust trends and trade patterns.

Fig. 8.4 Trade balances in median of *ambitious* scenarios by world regions 2030 and 2050. (**a**) solid biomass, (**b**) liquid. GFPM results are only included for 2030 and include only energy products based on forest biomass (Unit: Mt)

Fig. 8.4 (continued)

In ambitious scenarios, the key potential future bioenergy export regions in 2050 are Russia and former USSR countries (40 % of trade, 10 % of global demand) and Canada, South-America, Central and Rest Africa, Oceania (40 % of trade, 10 % of global demand). This general pattern also holds for the moderate scenarios with slightly shifted figures: Russia and former USSR (33 % of trade, 6 % of global demand), Canada, South-America, Central and Rest Africa, Oceania (60 % of trade, 12 % of global demand). For the USA, there is a significant difference in the trade balance of liquid vs. solid biomass. Where the scenarios show a quite balanced (or slightly positive) trade balance for solid biomass, the trade balance for liquid bioenergy is clearly negative.

Regarding the key future import regions in scenarios up to 2050, mainly India, Western Europe and China are dominating. In ambitious scenarios these three regions import more than two thirds of all global inter-regional trade: India (33 % of trade, 8 % of global demand), Western Europe, China (39 % of trade, 9 % of global demand). USA is a relevant importer of liquid biofuels, however this is partly compensated by exports for solid biomass. The moderate scenarios show a more balanced picture: India (42 % of trade, 8 % of global demand), Western Europe (33 % of trade, 4 % of global demand), several world regions holding a share of about 3–6 % of global trade and about 1 % of global demand, e.g. Japan, China, South-East Asia and Rest of South-Asia, Middle-East and North Africa, USA, Korea, Turkey. For India, the scenario results are in a very close range, whereas for China a high difference between model results can be observed. This indicates the substantial uncertainties regarding biomass potentials and future exploitation of these potentials in China.

In the long-term (i.e. after 2030), the scenarios show a declining demand for liquid biofuels in Europe and the USA which reduces the imports from these regions.

In particular, the results regarding the relevance of Asia as importing region are also supported by Raunikar et al (2010) as well as IEA 2012.

However, one should keep in mind that the trade flows identified above are from models that are in first instance not made to analyse bioenergy trade. They are simply a consequence of where the models predicts demand for and supply of biomass. When comparing the trends identified above with current actual trade flows, the following observations can be made:

- Russia and other former USSR countries, whilst possessing very large biomass resources, have so far only been a minor exporter of solid biomass, whilst trade in liquid biofuels is virtually non-existent. In between 2010 and 2012, wood pellet production capacities have been expanded strongly, especially in North-West Russia, but also in Russia's East (aiming to feed the East Asian markets) so this could indeed be a start of substantial solid biomass exports in the years to come.
- Canada has been one of the pioneers of solid biomass exports, and the expected major role as a biomass supplier fits current trends quite well.
- Latin America and Africa on the other hand virtually do not export any solid biomass at the moment, and are also not likely to do so in any significant volume until 2020. Thus, huge exports of solid biomass from these regions in the near

and mid-term future are rather unlikely. Significant barriers would have to be overcome and logistical, social, ecological and economic challenges would have to be solved. Exports of liquid biofuels from Latin America on the other hand are already significant (see Chap. 2), and could likely expand further in the decades to come. For Sub-Saharan Africa, which has experienced a number of failed biofuel projects in recent years, this still remains to be seen.

- One market (pending current and future policy developments) to increase its liquid and solid bioenergy imports further is the EU, as also the models anticipate. This is probably one of the most robust trends identified. The largest uncertainty perhaps are the future additional sourcing areas, i.e. if Latin America and the African West coast may become important suppliers in the future as well.
- To some extent, India and China still remain wild cards. Both countries have shown little or no bioenergy imports or exports so far, partly due to the lack of strong supporting policies stimulating demand and at the same time limited amounts of agricultural land and forests that could be used to produce biomass for energy. Both countries have large potentials of agricultural residues, but these are likely to be used locally. It remains to be seen, if the large bioenergy imports expected by the model results will materialize. If so, from a logistical point of view it would make sense if India and China might increasingly source biomass from the east-coast of sub-Saharan Africa, while China might also utilize the forest biomass in East Russia. However, both bioenergy trade routes are virtually non-existent today, and thus would have to be developed from scratch. Again, any scenarios (implicitly) expecting large trade flows following these routes in the short and mid-term should be considered with caution.

Finally, it should be pointed out that model outcomes of these trade flows depend on how a number of issues are included in the models. These include bioenergy availability and cost, bioenergy demand and bioenergy trade barriers and logistics. The details of how these trade-flows happen and what parameters drive/limit them should be investigated. Typically, when trying to mobilize such potentials, the initial costs are typically much higher than originally anticipated – also because significant cost reductions can usually only be obtained with increases in scale. Transaction costs for bioenergy trade are rarely included in global models, and if so are included in a crude manner. Underlying assumptions and sensitivities may significantly affect the results. It is therefore deemed worthwhile investigating if the costs of transport should also be modeled as a function of scale and cumulative production.

8.4.2 Future Challenges and Open Questions for International Bioenergy Trade

In this chapter, a comparative investigation was carried out of selected model scenarios regarding bioenergy demand, production and the implication on bioenergy trade between world regions. Those model scenarios with an ambitious increase of bioenergy demand imply a huge increase in bioenergy trade, an increase by a factor of 70 between 2010 and 2030 for liquid biofuels, and by a factor of 80 for solid biomass. It has to be taken into account that these results refer to trade between world regions. International trade within these regions (e.g. within Europe) would have to be added to these values. Such an increase would result in quantities of internationally traded biomass commodities which would be higher than the current total global bioenergy demand (i.e. larger than 50 EJ). Considering the currently very small share of internationally traded bioenergy, this would result in huge challenges and tremendous changes in terms of production, pretreatment of biomass and development of logistic chains. While both liquid and solid international biomass trade has grown exponentially between 2000 and 2010, it is rather doubtful that this speed can be maintained and reach the levels of trade anticipated by the models. As an illustration, worldwide coal trade amounted to 1,142 Mt in 2011 (world coal 2013), i.e. roughly the size that solid biomass would need to grow to within 20 years in the ambitious bioenergy use scenarios. However, coal infrastructures have been developed for over 200 years, coal does not require any pretreatment before transport, and logistics typically originate from large point sources (mines).

From the above, two conclusions can be drawn:

1. Current global energy models seem to overestimate the amounts of liquid and solid biomass that can be traded especially in the medium term (2030), as it would require extremely high annual growth rates, which could only be accommodated with very high investments in production facilities and logistic infrastructure. However, it should be taken into account that the models do not make predictions. They provide scenarios based on biophysical trends and observed historic behavior under certain conditions and assumptions. The models tell what is potentially possible. Their objective is not to give advice how to overcome certain barriers. So, one reason of this overestimation could be that barriers for trade are not sufficiently covered in the models. If this is true, the question arises: How would global scenarios change if bioenergy trade barriers would be taken into account? To which extent would this change our picture of future global bioenergy use? So far, only a few number of global energy models explicitly simulate international bioenergy trade. Nevertheless, all global energy scenarios need to make an assumption on the future development of bioenergy trade. Mostly, this is only implicitly the case and is not clearly documented. A further investigation and integration of international bioenergy trade, barriers and drivers into existing modeling frameworks is crucial for a proper understanding of bioenergy in the future energy system. We recommend that modellers investigate their model–specific assumptions and outcomes for international bioenergy trade, and analyse whether the required growth rates in international bioenergy trade can be deemed realistic. Also users of the model results (e.g. industry, and policy makers that follow the IPCC reports) should be made aware of these model limitations.
2. The level of international bioenergy trade shown in the model scenarios is necessary to fill the anticipated regional gap between demand and supply. Without significant bioenergy trade between world regions, a much less pronounced

growth of bioenergy is achievable. Hence, either major challenges regarding amongst other technical, logistical and economic aspects of international bioenergy trade will have to be solved, or the objectives of significant higher bioenergy use have to be reduced. Policy makers should thus realize that next to incentives to promote production and consumption of bioenergy also policies to deal with the (rapid) growth of bioenergy trade will need to be put in place.

The insight into future scenarios and perspectives of bioenergy trade revealed that substantial challenges for the future development of global and international bioenergy trade may be expected in the coming decades if a low carbon energy system is to be developed. Some of these, such as the development of logistics, the required investments to realize production and trade, and the need to govern sustainable production of bioenergy are addressed in this book. Others are still open for further research, e.g. the implications of bioenergy trade for specific regions and for different biomass commodities in terms of social, ecological and economic impacts or the effect of fluctuating exchange rates, regional development of economic and policy side conditions.

References

Berndes, G., Hoogwijk, M., & van den Broek, R. (2003). The contribution of biomass in the future global energy supply: A review of 17 studies. *Biomass and Bioenergy, 25*(1), 1–28.

Bouwman, A. F., Kram, T., & Goldewijk, K. K. (2006). *Integrated modelling of global environmental change. An overview of IMAGE 2.4*. Bilthoven: Netherlands Environmental Assessment Agency (MNP).

Buongiorno, J., Zhu, S., Zhang, D., Turner, J., & Tomberlin, D. (2003). *The global forest products model*. San Diego: Academic/Elsevier. 301 pp.

Buongiorno, J., Raunikar, R., & Zhu, S. (2011). Consequences of increasing bioenergy demand on wood and forests: An application of the global forest products model. *Journal of Forest Economics, 17*, 214–229.

Gielen, D., Fujino, J., Hashimoto, S., & Moriguchi, Y. (2003). Modeling of global biomass policies. *Biomass and Bioenergy, 25*, 177–195.

Hamelinck, C., & Hoogwijk, M. (2007). *Future scenarios for first and second generation biofuels* (p. 86). Utrecht: Ecofys.

Hansson, J., & Berndes, G. (2009). Future bioenergy trade in the EU: Modelling trading options from a cost-effectiveness perspective. *Journal of Cleaner Production, 17*(Supplement 1), S27–S36.

Havlík, P., Schneider, U. A., Schmid, E., Böttcher, H., Fritz, S., Skalský, R., Aoki, K., Cara, S. D., Kindermann, G., Kraxner, F., Leduc, S., McCallum, I., Mosnier, A., Sauer, T., & Obersteiner, M. (2011). Global land-use implications of first and second generation biofuel targets. *Energy Policy, 39*, 5690–5702.

Heinimö, J., Ojanen, V., & Kässi, T. (2008). Views on the international market for energy biomass in 2020: Results from a scenario study. *International Journal of Energy Sector Management, 2*, 547–569.

Hoefnagels, R., Junginger, M., Resch, G., & Panzer, C. (2011a). *Long term potentials and costs of RES. Part II: The role of International Biomass Trade*. A report compiled within the European research project RE-Shaping, August 2011. Available at: http://www.reshaping-res-policy.eu/downloads/WP5_ReportD12%20FINAL.pdf

Hoefnagels, R., Junginger, M., Resch, G, Matzenberger, J., Panzer, C., & Pelkmans, L. (2011b). *Development of a tool to model European biomass trade* (Report for IEA bioenergy task 40). Available at: www.bioenergytrade.org

IEA. (2012). *World energy outlook 2012*. Paris: International Energy Agency.

Ince, P. J., Kramp, A. D., Skog, K. E., Yoo, D.-i., & Sample, V. A. (2011). Modeling future U.S. forest sector market and trade impacts of expansion in wood energy consumption. *Journal of Forest Economics, 17*, 142–156.

IPCC. (2011). *IPCC special report on renewable energy sources and climate change mitigation*. Prepared by Working Group III of the Intergovernmental Panel on Climate Change (O. Edenhofer, R. Pichs-Madruga, Y. Sokona, K. Seyboth, P. Matschoss, S. Kadner, T. Zwickel, P. Eickemeier, G. Hansen, S. Schlömer, C. von Stechow, Eds.). Cambridge/New York: Cambridge University Press.

Kallio, A. M. I., Moiseyev, A., & Solberg, B. (2004). *The global forest sector model EFI-GTM – The model structure*. Joensuu: European Forest Institute.

Lamers, P., Hamelinck, C., Junginger, M., & Faaij, A. (2011). International bioenergy trade – A review of past developments in the liquid biofuel market. *Renewable and Sustainable Energy Reviews, 15*, 2655–2676.

Lundmark, R. (2010). European trade in forest products and fuels. *Journal of Forest Economics, 16*, 235–251.

Masera, O., Ghilardi, A., Drigo, R., & Angel Trossero, M. (2006). WISDOM: A GIS-based supply demand mapping tool for woodfuel management. *Biomass and Bioenergy, 30*, 618–637.

Matzenberger, J., Daioglou, V., Junginger, M., Keramidas, K., Kranzl, L., & Tromborg, E. (2013). *Future perspectives of international bioenergy trade* (Report of IEA bioenergy task 40). Available at: www.bioenergytrade.org

Muñoz, P., Giljum, S., & Roca, J. (2009). The raw material equivalents of international trade. *Journal of Industrial Ecology, 13*, 881–897.

OECD. (2012). *OECD environmental outlook to 2050: The consequences of inaction*. Paris: OECD.

Raunikar, R., Buongiorno, J., Turner, J. A., & Zhu, S. (2010). Global outlook for wood and forests with the bioenergy demand implied by scenarios of the Intergovernmental Panel on Climate Change. *Forest Policy and Economics, 12*, 48–56.

Smeets, E. M. W., Faaij, A., et al. (2007). A bottom-up assessment and review of global bio-energy potentials to 2050. *Progress in Energy and Combustion Science, 33*(1), 56–106.

Sokhansanj, S., Kumar, A., & Turhollow, A. (2006). Development and implementation of integrated biomass supply analysis and logistics model (IBSAL). *Biomass and Bioenergy, 30*, 838–847.

Szabó, L., Soria, A., Forsström, J., Keränen, J. T., & Hytönen, E. (2009). A world model of the pulp and paper industry: Demand, energy consumption and emission scenarios to 2030. *Environmental Science & Policy, 12*, 257–269.

van Vuuren, D., Den Elzen, M. G. J., et al. (2007). Stabilizing greenhouse gas concentrations at low levels: An assessment of reduction strategies and costs. *Climatic Change, 81*(2), 119–159.

World Coal. (2013). *Coal market & transportation*. Fact sheet by the world coal association. Available at: http://www.worldcoal.org/coal/market-amp-transportation/

Yamamoto, H., Fujino, J., & Yamaji, K. (2001). Evaluation of bioenergy potential with a multi-regional global-land-use-and-energy model. *Biomass and Bioenergy, 21*, 185–203.

Chapter 9
Financing Bioenergy Trade: Making It Happen

Michael Deutmeyer, Bo Hektor, and Peter-Paul Schouwenberg

Abstract The development of sustainable biomass supply chains for international biomass trade are a prerequisite to foster the growth of bioenergy applications worldwide. Setting up these supply chains is offering a broad array of interesting investment opportunities that could offer stable, long term and high returns. However, investment in the bioenergy supply chains is perceived by the finance sector to be risky and uncertain and therefore many projects do not materialize. At present, insufficient financing is an obvious obstacle for the development of efficient supply chains. In the chapter attempts are made to analyse and elucidate the causes for that perception in the finance sector and to suggest remedies. This chapter also highlights the different areas of investment that exist, the most important stakeholders along the investment process, the challenges that exist along certain bioenergy value chains and the need for far sighted and sound policy making to support and secure long term sustainable business models.

9.1 Introduction

The focus of this chapter is on investments needed that enhance the international trade of sustainable biomass.

During the first decade of the twenty-first century about 10 % of worldwide primary energy demand has been met with biomass based fuels (IEA 2007).

M. Deutmeyer (✉)
Green Resources, London, UK
e-mail: michael.deutmeyer@greenresources.no

B. Hektor
SVEBIO, Stockholm, Sweden

P.-P. Schouwenberg
RWE Essent, s-Hertogenbosch, The Netherlands

M. Junginger et al. (eds.), *International Bioenergy Trade: History, status & outlook on securing sustainable bioenergy supply, demand and markets*, Lecture Notes in Energy 17, DOI 10.1007/978-94-007-6982-3_9, © Springer Science+Business Media Dordrecht 2014

Already more than 10 million tonnes of solid and liquid biofuels are being traded internationally per annum (Junginger et al. 2010) and according to the estimates of AEBIOM (2012) overall bioenergy consumption in Europe alone will grow from 100 million toe in 2010 to more than 120 million toe in 2020. An increasing share will be traded internationally in order to balance demand and supply and/or optimize costs.

The development of sustainable biomass supply chains for international trade is often capital intensive, technologically demanding and time consuming. However, without a stable and reliable supply of sustainably produced biomass not all bioenergy projects needed to meet renewable energy mandates can be realized or have to revert to unsustainable feedstock sources. It is believed that market forces alone will not suffice to balance supply and demand of biomass on a world wide scale – also due to different incentive schemes and trade distortions. Also politics have to play an active role in managing and guiding the development of biomass supply chains.

Biomass to energy value chains – especially when involving long haul biomass supply chains – are far more complex to establish and to manage than other renewable energy systems. Biomass availability might be volatile or seasonal and comes in a huge variety of specifications, there is a broad range of players involved from single farmers or land holders, project developers, private and institutional investors, transportations companies, traders up to internationally operating utilities. All with different agendas, ways of doing business and bargaining powers.

In addition different preconditioning technologies offer a broad range of intermediate biomass energy carriers – most of them requiring a specific conversion technology to convert them at optimum efficiency to power or fuel.

Finally the low liquidity of the international biomass market does not allow for large scale investments in e.g. biomass power plants without securing a significant amount of feedstock via long term contracts – leading to totally integrated and complex supply chains.

As a result the investments in international biomass supply chains are currently not sufficient to meet future renewable energy obligations. Long term trends of energy markets as well as supportive political and economic frameworks already today allow for robust business models in that investment class.

A proactive approach is needed by the bioenergy community in order to provide the financial world with information about potentially lucrative investment opportunities in biomass trade.

This chapter should therefore deal with the following questions:

- What kind of attractive business models exist along the biomass trade supply chain?
- What kind of investments are already being developed, where is a lack of activity today?
- Who will finance biomass trade related investments and business models?
- What kind of financial sources are available for those potential investors (grant, debt, equity)?
- How to align and synchronize renewable energy mandates with needed investments?

Concluding remarks will try to answer the questions of what could be done in order to support the growth of international biomass trade and what kind of economic or political incentives might still be missing.

9.2 International Biomass Trade and Investments

9.2.1 Biomass Supply Chains of International Biomass Trade

International imbalances of biomass supply and demand as well as biomass costs are the drivers of a growing international trade in biomass. International biomass supply chains differ in a number of points from local or national biomass supply chains that both add to their complexity but also offer new and exciting business and investment opportunities.

One important and mainly logistical aspect of those supply chain is the question of both the technical and economical ***transportability*** of the solid or liquid biofuel in question. Quite a few of sub criteria have to be fulfilled to have a transportable biomass product in that sense.

One of the criteria is ***volume***. Only if a certain biomass fuel can be supplied in sufficient total volumes per year in order to allow for transports by e.g. sea going vessels a regular business can be established that both reduces overall costs and attracts investors. This is especially the case if specialized handling, storage or transport equipment might be needed along the supply chain. Only if whole bulk carriers can regularly be filled with e.g. energy pellets or energy wood chips of a well-defined quality international trade has a chance of getting started.

Especially the dedicated production of energy biomass can be a solution to securing high and continuous volumes of biomass. Here biomass is the target product and no longer a side product of other production chains and no longer depends on other industry cycles. In the case of dedicated biomass production often new plant traits and growing regimes are being developed in order to maximize overall energy content. Short rotation coppice with high yielding tree species, new and cost efficient tree planting and harvesting technologies as well as high biomass sorghum or maize plants are examples of new market and investment potentials mainly driven by an increasing bioenergy demand.

Creating the necessary liquidity of biomass in certain regions to establish international biomass trade is a business model in itself and has happened for example in British Columbia more than a decade ago. Nowadays there are more than 11 wood pellet plants operating in BC with a total capacity of more than 1.7 million tons of pellets per year – all destined for mostly European export markets.

Total internationally traded pellet volumes are around 8–10 million tons while soon the import to Europe alone has to rise to about 25–50 million tons during the next years to bridge the gap between expected demand and European supply.

In addition to scale also transport costs as such have to be optimized. Therefore the biomass fuel needs to have good ***storage*** and ***handling*** properties as well as a high volumetric ***energy density***. Often biomass feedstock does not have some or most of these properties. That leads to the need of some form of technical pre-treatment prior to long haul shipments – a fact that led to the growth of a whole new set of biomass pre-treatment technologies during the last two decades such as wood pelletisation, torrefaction, flash pyrolysis, hydrothermal carbonization, enzymatic hydrolysis, etc. Wet wood chips have an energy density of about 8 GJ/t or 3 GJ/m^3 in comparison to pellets with 17 GJ/t and even 11 GJ/m^3. All these technologies are at different stages of their development and commercialization and offer a multitude of investment opportunities both on the development as well as on the application of these technologies.

International trade also needs a competitive and regulated environment in order to flourish without harming the environment. Biomass fuels traded worldwide need to follow common ***specifications*** in order to become commoditized to be easily tradable and allow for the development of financial instruments to manage risk of long and medium term supply agreements that in turn are the basis for investments needed along the total supply chain up to the end user. Besides the ***commoditization*** of biomass also the ecological and social ***sustainability*** of such fuels have to be guaranteed via specialized certification systems. Especially for exports to the EU – currently the biggest importer of solid and liquid biofuels, such certifications are compulsory. Therefore investments in easily certifiable and sustainable biomass resources could become very lucrative as demand for those qualities will rise in the future.

All these investment opportunities need to find the right source of finance to bridge the gap between project idea and project realization. Depending on the degree of technical and economic risk, the degree of innovation and IP generation as well as the degree of future income potential and market position, different portfolios of financial sources and financial tools are or could be made available.

9.2.2 Short Overview of Investment Activities in International Biomass Trade, Drivers and Future Bottlenecks

Main areas of investment along international biomass supply chains are the fields of dedicated biomass production, ***biomass preconditioning***, ***biomass logistics*** and ***biomass conversion***.

9.2.2.1 Production

With the increasing need for bioenergy fuels existing sources of biomass residues and waste do no longer suffice to meet demand or specifications. There are only a few biomass residues that are being produced on a continuous basis and in large quantities that have all the needed properties as defined in the previous chapter to be directly suitable for international trade such as palm kernel shells in South East Asia.

In general dedicated biomass production will be the very starting point of most biomass value chains ensuring security of supply, absolute overall volumes for international logistics and continuous feedstock quality to meet specifications. A number of recent investment in pellet plants to supply oversee markets are based on such a feedstock base.

Dedicated production of woody energy biomass for export is still in its infancy and so far has only been started to serve local or regional biomass markets, such as short rotation coppice (SRC) in Europe – with a couple of specialized companies (e.g. Lignovis, Agroenergi) active in that field already. Total area covered by SRC in Europe is estimated to be as little as 18,000 ha although widely recognized as one of the future sources of sustainable biomass sources that easily competes with imports when measured in price per GJ delivered to the gate. The usage of existing pulp wood plantations in e.g. South East USA for the production of wood pellets via whole tree chipping could be interpreted as dedicated biomass production activities for bioenergy exports. Also in countries such as Brazil, South Africa and Australia energy wood plantations are being developed to cater for both local and international bioenergy markets.

In the case of liquid biofuel production already existing agricultural production technologies and concepts have been used to increase the production of sugar, starch and oil crops such as sugar cane in Brazil, maize in the USA, soya in Argentina and palm oil in Malaysia in order to divert significant volumes into the energy market.

Lignocellulosic biomass, however, can be produced very extensively with low external energy input. Especially in the case of e.g. reforestation projects its overall GHG balance can be very positive with a much lower – in some cases even beneficial – impact on biodiversity in comparison to high intensive agriculture.

These activities in dedicated biomass production need to be developed and managed sustainably to make sure that the fight against increasing GHG emissions through bioenergy use will not be a cause of other unnecessary environmental problems and distortions. So when thinking about a new project for dedicated biomass production it is important that investors make sure that the project will meet the criteria of the European sustainability legislation and can be certified accordingly. Only a certified biomass production is eligible to enter the important and growing European biomass market. Certification will also be a precondition for quite a number of financial sources such as grants and institutional investors and greatly enhances the chance of success and continuous growth.

9.2.2.2 Preconditioning

Biomass needs to be preconditioned to meet technical specifications and enhance its storage, handling and transport properties for long haul transport chains.

Pelletisation is currently the most ubiquitous form of solid biomass preconditioning for that purpose. Significant investments in pellet plants have been done during the last ten years in order to make low cost biomass and biomass residues available for international markets. With Europe as the main importing nation, numerous pellet plants have been set in Western Canada, South East USA and Russia, almost totally geared towards the international biomass trade. Similar initiatives in South Africa and Australia have been seen but with limited economic success and low capacity utilization rates so far. The latest IEA task 40 pellet study (Cocchi et al. 2011) offers a good overview of the development and current status of the world wide wood pellet industry that had a total capacity of about 28 million t/a and a total output of about 14 million t/a in 2010. About 5–6 million tons of pellets have been traded on international routes the same year.

Another already commercially viable and implemented biomass preconditioning technology for international sales is wood chipping and drying with established supply chains based in e.g. in the South of the US for supplying dedicated biomass power plants in Europe. Such projects could be an ideal starting point in countries with a high surplus of low quality and sustainable energy wood and simple transport chains (few handling steps) to reach international markets. The mobilization of unproductive while over aged rubber trees from Liberian tree plantations – that needed to be cut down prior to replanting – to the international biomass market through Buchanan Renewable just made use of relatively low cost whole tree chipping and truck/ship transport.

Impressive amounts of investments were done also in biodiesel plants and ethanol plants to serve the fast increasing bio fuel market during the last decade. Those investments were concentrated in EU countries and the Americas (mainly USA, Brazil, Argentina). While some of the produce was also geared to international biofuel markets, already existing storage, handling and transport infrastructure for fossil transportation fuels provided sufficient capacity for those volumes to be transported at optimized costs to their respective destinations. With increasing biofuel consumption and based on well-established feedstock and transportation base, investments in biofuel plants have been very lucrative during the last decade. The projections by the OECD and FAO indicate that global production of ethanol will double to 125 billion litres by 2017 with biodiesel production rising from 11 billion litres in 2007 to about 24 billion litres in 2017.

The investment focus in the field of liquid biofuels should now be directed to second generation biofuels that are based on a new and improved feedstock base such as biomass waste or high energy crops instead of sugar, oil or starch crops and new conversion technologies such as enzymatic hydrolysis and biomass gasification cum synthesis technologies (CHOREN, Range fuels, Coskata, etc.). Here a totally new industry opens up that will also be able to supply a future biomass based chemical industry with its most essential intermediary molecules and building blocks.

Also considerable investments took place in the development of new preconditioning technologies of solid biomass such as torrefaction (Deutmeyer et al. 2012), flash pyrolysis and hydrothermal carbonization. First plants for lignocellulosic biomass are already operational in North America and Europe as well as first commercial torrefaction plants (New Biomass Energy/USA, Thermya/France, etc.) (Deutmeyer et al. 2012).

9.2.2.3 Transportation and Handling

There are also a number of investment opportunities to be found in biomass logistics. Specialized rail wagons or high cube containers (such as e.g. "Innofreight" containers) to carry low density biomass fuels such as pellets or wood chips are increasingly in use and offer sound business models. Also specialized biomass hubs such as the Fibreco biomass terminal at Vancouver (BC) experience solid growth rates and profits for their investors. Also the European ports of Antwerp, Amsterdam and Rotterdam are getting prepared for an increase in biomass imports to the EU.

Although there seems to be no current shortage in long haul bulk carriers the focus of investments in biomass logistics lies in adequate storage and handling facilities that allow to store huge volumes of biomass fuels at the port and efficiently load these onto bulk carriers in order to reduce overall charter costs. The same holds true for the receiving ports that need to discharge, storage and transship biomass fuels in an equally efficient and low cost manner.

9.2.3 Basic Considerations for Investments

9.2.3.1 Biomass Demand

The greatest demand for imported biomass during the next decade will be from Europe, Korea and Japan. In ***Europe*** the "RE 20/20/20" energy policy carries legally binding renewable energy targets for each member country for 2020. Plans submitted by member countries in 2010 to achieve targets will increase biomass use for production of electricity, heat, and transportation fuels by ~400 MT (million tonnes), mostly based on woody biomass feedstock.

Pellet consumption of 9 MT in 2009 is projected to reach 16–18 MT by 2013–2015 and 50–80 MT by 2020. While Europe could meet the rising biomass demand by increased harvesting of her forests and from energy plantations, an increase in biomass imports is more likely since in most cases less costly.

The European biomass shortfall is estimated at 60 MT and leading to a high growth potential for biomass imports. Key importing countries will be UK, Netherlands, Belgium, Germany, Italy and Spain. In ***North America***, the US and Canada do not have ambitious renewable energy plans, though support has been given for specific purposes; e.g. the US supports producing ethanol from corn for transportation fuel, but mainly to reduce dependence on Middle East oil and provide income to farmers. In Canada renewable fuel targets are low and on the verge of being achieved. An extensive natural gas distribution system makes the economics of bio-heat and biomass power difficult. Canada is expected to remain a net exporter of biomass to 2020. China's new 5-year plan focuses on renewable energy. Domestic demand will increase substantially but it supports self-sufficiency and biomass trade is not yet envisioned. Korea has an ambitious target for renewable energy and a large scale import program (15 MT of pellets for co-firing). In post-tsunami Japan, massive domestic and import biomass programs are contemplated that would increase demand for imported biomass to the same magnitude as Korea. Overall biomass import demand of these three key importing regions could add up to 90 MT by 2020 alone.

9.2.3.2 Biomass Supply

Globally, solid biomass is mostly traded in an open market, supported by energy policy. In 2009, 30 % of 13 MT of wood pellets produced were exported, primarily from Canada and the US. Bioethanol and biodiesel are subject to import tariffs and other restrictions, so even though biodiesel trade grew 0–80 PJ in 2005–2009 only 14 % was exported, and only 3 % of ethanol, 95 % from Brazil and the US. A growing portion of biomass for liquid fuels will be imported. Liquid biofuels will be from by-products of established industries, sugar, palm oil etc. Solid biofuels such as pellets, traditionally sourced from sawmill residues, will increasingly come from more intensive utilization of industrial, forestry and agricultural residues, inferior trees, and forests destroyed by fire, storm and insects. Canada has approximately 50 MT of excess biomass annually in mill residues, hog piles, unutilized allowable cut, non-merchantable timber and urban wood, and an almost unlimited wood supply from insect infested BC forests. Brazil has excess fibre from its forest industry, and potentially 25 MT of unused sugar cane bagasse. The US has large potential in the South-Eastern States. In the long term, supply will move towards biomass plantations on abandoned or under-used land in superior growing areas, and new regions. Coastal Africa has large tracts of diseased or invasive wood species. Argentina has untouched mill residues. Large land potentials are seen in South America, Africa, Australia, and parts of Asia. Plantations will require attention to ecological, political and cultural issues to avoid impacts on food and fodder production. Long term supply contracts will persist, but some will be traded as a commodities.

9.2.3.3 Bioenergy Risk

Like any investment, bioenergy has to deal with a number of risk. The most important risk to mitigate is the ***supply of biomass*** feedstock. Wind turbines are regarded as safe investments, though wind availability is uncontrollable. Biomass supply is controllable, storable, and can be guaranteed if based on dedicated production. Sawmill residue is low-cost, homogeneous, and often close-by, but supply depends on long term viability of the sawmill. Harvest residues may be plentiful, but often are not homogeneous, and supply systems are complex. Standing timber is often not owned, but available under licenses or market conditions. Wood plantations provide for long-term supply, but are at risk to sustainability issues. Agricultural residues are usually available only for a short period after harvest. Any bioenergy project must accept complex agreements to limit biomass supply risk as much as possible. Bioenergy is perceived to have high ***technology risk*** by investors, but that is only the case with new non-commercial technologies. Most bioenergy projects have low technology risk; they use proven processes and equipment and have well understood supply chains. Pellet manufacturing is well known, whereas torrefying wood and subsequently pelletizing is still in its infancy. Worldwide 3–4 companies claim they are producing at commercial scale, but most are still at the pilot stage. Only one company has proven consistent production of pyrolysis oil at the commercial scale, while a second has produced for lengthy runs, but not over several years. ***Transportation risk*** can be significant, evidenced by the volatility in maritime shipping prices 2006–2009 due to demand for shipping by the Chinese economy. Such risk can be mitigated by long term shipping contracts, or dedicated specialized ships. ***Regulatory risk*** is a major factor. The economics of renewable energy, including wind solar and bioenergy, is supported by government incentives and renewable portfolio standards. Occasionally government policies may be reversed, evidenced by the withdrawal of feed-in-tariffs in the Netherlands. The safest markets are those in which economics of bioenergy are still acceptable in a regulation-free market. ***Market risk*** can be a factor. Pyrolysis oil is a very dense energy medium, but it is a new product not well known by potential markets. Bioenergy data is often scarce and public information poor, professional education and training is in its infancy. Like wind power, bioenergy is vulnerable to disinformation and adverse lobbying by competing stakeholders and by a misinformed environmental community. ***Counterparty risk***, the risk of either party not being able to live up to its contractual obligations, can be significant, as in any business. For example, sawmill closures in Western Canada forced pellet suppliers to reduce production due to shortfalls in mill reside supply, and requiring pellet mills to find new sources of fibre. Pellet buyers had to secure volumes from an illiquid pellet market at high prices.

9.2.3.4 Country Risk

Countries with surplus biomass such as Canada, New Zealand and the US are viewed as low risk targets for investment due to stable political systems, western financial systems, solid infrastructure etc. Many other regions have surplus biomass and no meaningful domestic demand but are considered high risk. Tropical countries in East and West Africa, South East Asia and Latin America have very good growing conditions and under-utilized agricultural land, but for many European investors many of these regions are not on their radar screen. For biomass investments to be successful it is critical to assess and mitigate country risk factors in regions selected for investments. Individual country risk profiles should be developed, including political, legal, economic and cultural aspects. Quantitative country scorings and qualitative risk indicators can be used. Asset risks can be mitigated by developing modular plants or acquiring loading equipment that can be dismantled and moved if necessary. Regional risk can be further reduced by involving international and national financial or development institutions in investments, such as the IFC, World Bank, KfW, DEG, and participation by reliable local partners that have lengthy business experience in that country. Win-win situations for both the fund and the regional population greatly enhance the security of the investment.

9.2.3.5 Investment Selection Criteria

It is the intent of any investment to make a satisfactory or superior return commensurate with risk. Investments in biomass value chains intent to maximize biomass trade; developing biomass in low-cost regions and profitably enabling long-distance transport with a landed cost lower than the market can supply internally. It might consider investment selection criteria as follows; maximize biomass supplied to markets, acceptable rate of return for risk, 50 % of investments in a positive cash flow, <30 % of investment in high risk regions, provide jobs in plant locations with a reliable work force, secured sustainable low-cost biomass feedstock base, secured off take agreements, 80 % proven technology, world scale facilities with competitive costs, low exchange rate risk, access to international shipping, high potential for efficient ground supply chain logistics.

9.3 Investors in Biomass Trade Supply Chains

9.3.1 Government

The role of governments as investors in international biomass supply chains are mainly restricted to direct investments in general infrastructure projects such as rail ways, transportation roads. These forms of investment support the possibility of the respective country to engage in the international trade of bioenergy products but in most of the cases are serve a broader range of development general goals.

More specific support for enhancing biomass trade is given more indirectly through the creation of suitable import and export legislation for bioenergy products, support in the development of new technologies for the production of specification driven bioenergy carriers, mandates for the use of bioenergy and renewable energy primarily in the transportation, heat and power sector and through direct subsidies in individual investments along the biomass to energy value chain – starting from subsidizing the establishment of dedicated biomass plantations up to the support in investment in bioenergy conversion facilities such as pellet stoves or biomass power plants.

9.3.2 Industry

The most important industrial investor into international biomass supply chains are European utilities that need to secure the needed solid biomass feedstock for current and future bioenergy mandates. They had realized that in an underdeveloped international biofuel market with considerable long term price risks (and chances for private investors) a full integration into the biomass to energy value chain offers them a couple of advantages that set of the higher complexity of their bioenergy business model.

Companies such as e.g. RWE Innogy and Drax Power now have started to secure low cost woody biomass feedstock in Southern USA and are investing in large scale pellet plants (Waycross by RWE Innogy at 900,000 t/a; Bastrop/Gloster by Drax Power at two times 450,000 tpa, etc.) and port facilities (mainly large scale storage and efficient ship loading technology).

These investments will eventually pay off nicely since they will make use of the bioenergy base load of their investors and find a continuous and uninterrupted off take while at the same time providing essential insight information on the upstream part of the bioenergy value chain that are important to assess third parties' medium and long term contractual biomass supply offers.

Other investments, such as the one by Hafslund, a Norwegian power company, in a pellet plant (BioWood Norway), followed a different approach excluding the close geographical and economic integration into the feedstock source and just stopped operation before ever reaching full capacity. So pure size is not always a guarantor for a sound business model.

In addition to these effects these investments also increase market liquidity and will accelerated the process of commoditization of internationally traded bioenergy products. As soon as this phase has started those utilities will surely step out again and keep concentrating on their energy conversion and distribution business.

Individual utilities also started to support single technological developments such as direct investments in specific technology start-ups in order to see an early availability of advances in biomass preconditioning and conversion technologies. The investment of RWE Innogy's venture fund in a torrefaction start up (Topell) or sterling engine developer (Sterling DK) are just an example of such direct investments.

Similar investments are taking place in the bio fuel sector for similar reasons. Big Oil, agri-processors (such as ADM) and agricultural producers are being seen investing in both production facilities of first generation and second generation bio fuel plants as well as in biomass production and logistics.

The investment of international companies into bio fuel projects are numerous such as BP's total of seven billion USD investment in alternative energy since 2005. Also ExxonMobil is spending 600 million USD on a 10-year effort to turn algae into oil and Royal Dutch Shell has invested billions of dollars in a Brazilian bio fuels venture, buying up sugar cane mills, plantations, and refineries to make ethanol. In the U.S., Shell produces small lots of so-called drop-in bio fuels—engine-ready products that can replace gasoline—from a pilot plant in Houston that uses sugar beets and crop waste and had investments into CHOREN (a former German BTL technology company) or Iogen, a Canadian based second generation ethanol producer.

Although not all investments have been a success and not all announced projects finally had been realized, international energy companies are still the largest single group of investors into bioenergy and international bioenergy supply chains.

9.3.3 Institutional Investors

Institutional investors are generally risk averse and focussing more on the long term maintenance of value than quick and high gains. At the same time they can and only will offer significant amounts of financial resources for single projects and would be the ideal partners for investments with a low perceived risk and high upfront investment. The purchase of large areas of land for e.g. dedicated biomass production would be such a case where the institutional investor takes ownership of the land renting it out to an operating entity that takes all the risk and potential gains from growing energy crops on it. Also large biomass conversion facilities based on bankable technical performance guarantees, long term biomass supply agreements and feed in tariffs and run by an experienced operating company could be another example for suitable investment opportunities for this kind of investor.

Interested private or institutional investors in this investment class could also delegate the work of project evaluation, project selection and development to specialized bioenergy funds that focuses its activities along the biomass to energy value chain.

9.3.4 Private Investors

Private investors – due to their own restricted capital resources and restricted access to other capital markets – tend to focus on single and easy scalable investments along the biomass supply chain starting from dedicated biomass production such as farmers or forest owners to mostly biomass collection and biomass transport. Only if supported by venture capital or if combined with existing operations private investors venture into biomass preconditioning such as setting up a pellet plant adjacent to their saw mill. Quite a few private investors can be seen investing in initial stages of technology development almost always facing the problem to find additional investors to support the construction of demonstration plants and the commercialization of the technology.

9.4 Financing Investments in Biomass Supply Chains

9.4.1 Grants and Subsidies

Feed in tariffs, biofuel quotas, production obligations and other forms of subsidies are all geared towards increasing the investment in bioenergy use and bioenergy production and certainly do have direct and indirect effects on investments into the bioenergy sector as well as into international bioenergy supply chains. Besides increasing the price for some bioenergy products it is mostly the increase of absolute demand or total volumes that have the biggest effect on the growth of international bioenergy trade.

In many cases bioenergy mandates or demand cannot be satisfied at sustainable price level through local or national production so that low cost and/or underutilized biomass resources from other regions in the world can be used to flatten the supply curve and stabilize both prices and volumes needed.

In addition to these subsidies there are possibilities to cover part of the needed investment into biomass supply chains by direct grants. Investments in infrastructure, technology, forestry and agriculture especially in developing countries are often eligible for grant financing of up to 60 % of total investment costs.

A number of donor agencies and development banks provide for such assistance in order to foster sustainable forestry and agriculture, international trade, technology transfer and introduction of renewable energy.

Also in the area of bioenergy both for local consumption as well as for export quite a number of grants are available (PSI, EU ACP, NCF, EEP, KfW, etc.). Investments into the production of dedicated and sustainable biomass production can also be set up as CDM projects so be co-financed by selling up front their CERs after project validation. Also the costs of validation and CDM project development can be financed by specialized grants.

Experience in the sustainable use of grant financing has shown that grants should primarily be used to support investments that are basically economically viable without any grant once operational but do need grant support in order to cover perceived country risks and introduce new technologies, know-how and investors into underdeveloped sectors. It is this specific aspect where the often required 'additionality' of grant applications have to be asserted.

Local and regional development banks (see e.g. www.iadb.org, www.adb.org, www.afdb.org) are usually a good source of information – besides well-known internet search engines – to start looking for investment grants. But also investments in more developed countries are eligible for direct financial support. For example the American Recovery and Reinvestment Act, or the stimulus package, has allocated billions to renewable energy, and biomass is one of the key focus areas. By leveraging these grant, loan guarantee and tax incentive programs, it is possible to get a well-planned and organized biomass project funded today in the USA (see e.g.: www.grants.gov, www.dsireusa.org, etc.). These government programs are geared towards bridging the gap in funding biomass projects until the industry matures and traditional capital becomes available.

Also European governments are very supportive in the area of biomass conversion technology development and offer a wide array of supportive schemes and grants especially for the initial stages of product development. It is most often commercialization step of new technologies that are the most difficult to finance. They are often too advanced to be eligible anymore for government grants and still too risky for private investors to step in.

9.4.2 Debt

Another source of capital is bank financing via debt. Here it is important for the project developer to meet and satisfy the "Bankability Requirements" for financial institutions to make available debt financing for the construction and implementation of a biomass project.

All major project risks have to be identified and their effect on the project economics and its ability to repay the loan need to be understood and discussed. For a bioenergy project the most important risks factors as discussed in Sect. 9.2.3 are usually:

- forward price risk
- biomass availability risk
- project completion risk
- long term demand risk
- transportation and logistics risk
- new technology risk
- sustainability criteria risk (or the absence of applicable criteria), etc.

The project promoter needs either to allocate these risks to third parties via contractual arrangements or at least greatly mitigate these risks via intelligent hedging, conservative modelling, greater project flexibility, knowledgeable investors and supportive business partners.

Besides the above mentioned development banks, debt could be provided by any kind of bank interested in well-structured and sound bioenergy investments. However, banks specialized on renewable energy and/or development projects tend to have a better understanding and assessment of the different risks incurred by a specific project proposal and can come up with both better advise for the project promoter how to mitigate those risks and better financing conditions.

9.4.3 Private Equity

A significant share of equity financing of generally between 25 and 50 %, is the precondition of any of the above mentioned sources of financing – such as senior debt – to become available. Equity finance bears most if not all of the risks involved in any kind of investment and the more sophisticated equity providers will focus on these same risks as the banks do as part of their pre-money due diligence.

Securing equity financing is therefore the biggest obstacle for any project developer. Sources of equity can be local investors, strategic partners, equity funds and tax equity.

The ability for biomass-to-electricity companies to claim investment tax credits, production tax credits, bonus depreciation and accelerated depreciation opens up the door for tax equity investors for examples for US based investments or in any other country were similar support is provided. With a declining tax equity market e.g. in the US during the last years due to the general unfavourable economic environment it is now possible to get a grant from the federal government instead of the 30 % investment tax credit for certain biomass projects.

Most valuable sources of equity are those of either from strategic partners that at the same time will support the investment as e.g. future suppliers or off takers or from knowledgeable investors and industry insiders that will not only provide the needed capital during project initiation and development but be a constant source of industry intelligence and advise during continuous operation. In addition, having successfully secured equity finance can in most cases be seen as a prove of project quality and economic viability and alleviates the receipt of additional needed debt financing and grant money.

Although equity finance eventually asks for the highest rewards it comes at no cost for the project unless profits are being made and other financing costs have been served.

9.5 Synchronize Investments with Existing Mandates and Future Biomass Demand

9.5.1 Possible Effects of Under Investments

As mentioned above, general observations have resulted in the conclusion that investment in the field of international trade of bioenergy meet obstacles and difficulties with regard to raising capital (Bradley et al. 2010). The reasons relate to several factors: uncertainty, lack of statistics and performance data, institutional issues, political and currency risks, etc.

On the general level the "under-investment" leads to limitations of international trade and therefore in most cases a reduction to the access to cheaper bioenergy on the international market. That in turn favours local producers because of less competition, while it increases prices for end consumers. In national or local markets where bioenergy is subject to subsidies, limited imports of low cost bioenergy put an extra burden on public budgets and on tax-payers.

Looking at more specific effects of under-investment, one should distinguish between different supply chains. Ethanol and vegetable oils, and residues like PKM, PKS (palm kernel shells), etc. are traded in well-established forms in or parallel to the trade of food and fodder products emanating from the same production processes. Under-investment problems are less pronounced for these supply chains than for other bioenergy, and could be absorbed by the capacity of the trade actors, who normally are big companies in all the links of the supply chains. The major problems for international trade in this segment relate to political and environmental rules and regulations.

Thus, under-investment is found mainly for newly developed trade patterns and for new bioenergy fuels such as wood pellets, torrefied biomass products, pyrolysis oil, etc. Below, some effects of under-investment and other financial restrictions are listed.

For several years after the EU targets for 2020 were decided upon, ***very little investment in new combustion capacity*** took place in the European power utility sector. Apparently, in these years financial subsidies were too low to support investment in new bioenergy based generating capacity. As the utilities have a key role in the transformation of the energy sector to reach the 2020 targets, some national governments have increased the financial support to levels where utilities start to act, leading to large scale and fast investments in bioenergy capacity . These stop and go decisions have had negative effects on actors in the supply chains, and have also contributed to the uncertainty regarding bioenergy trade in within the financing community.

Production projects aiming at long distance supply of bioenergy can be classified into three main categories:

- Stand-alone projects
- New products in large corporations
- Upstream project for energy utilities

Regarding the effects of under-investment, the following points could be made:

- A typical ***stand-alone project*** for bioenergy has to rely on the general finance market for most of the capital. The project developer than meets the arguments from the financing bodies that the uncertainty should be reduced or absorbed by firm agreements for long term off-take, raw material supply, technology guarantees, shipping, etc. This puts the project developer in a very weak negotiation position vis-à-vis these agreement providers and the required commitments would be disadvantageous, sometimes far from reasonable market conditions. At this point of the process, several projects have been cancelled. Among the remaining projects, only few have survived under the original conditions.
- A number of bioenergy project have been developed as a ***new production line*** complementing the product programs of established corporations. Examples can be taken from the technology, forestry and shipping businesses. Financing in these cases has been easier than for stand-alone projects, especially in cases when the corporation possesses a good reputation and a solid balance sheet. Still in several projects under-investment has been at hand, mainly regarding investment in marketing, raw material acquisition and management organization. The fact that bioenergy trade still is immature and is lacking norms and institutions means that more attention should be given to soft project functions.
- In recent years several examples of ***up-stream investment by energy utilities***, etc. have taken place, e.g. for pellets and ethanol. Many of these projects are large scale investments focusing on captive supply of bioenergy to own facilities. As most of these utilities are well established in the financial markets, and the fact that the off-take is guaranteed by internal captive use, financing can be arranged within established structures. In addition, many of the captive large scale projects are supported by public funds, projects in developing countries from foreign aid

funds. Moreover, export projects located in the US benefit from support from state investment funds, and also from subsidies under the federal Farm Bill.

Thus, the financing problems in bioenergy leads not only to general "under-investment" but also to the emergence of a structure in which the large utilities have the power to set the rules, to influence the pace, and to develop institutions for the bioenergy trade. Other equally sized sectors in the bioenergy fields, e.g. the small and medium scale heat markets, small fuel producers, independent agents and distributors will suffer more from limitations of finance and have great problems in developing roles for themselves in the growing international trade of bioenergy.

9.5.2 Mandates and Financing

In several cases, mandates are set up to enhance Renewable Energy. The 20/20/20 target for the EU is one example; similar schemes have been launched in Korea, Japan and in other countries.

The policy base for the mandates have been developed and decided upon on political levels. Thus, the mandates comprise several political goals in addition to environmental and renewability issues also jobs, innovations and increased self-sustainability are included. The financing of the implementation of the mandates reflects the political targets, however, the rules and regulations show a wide array of administrative arrangements, from support directly from the national budget to compulsory quota, etc.

In most cases, bioenergy is regarded as a key element in the policy. However, even in cases where it is obvious that the mandated target would lead to large scale import of bioenergy, very little if any attention has been given to activities aiming at the development of the efficiency of the supply chains of international trade. This approach may partly be explained by the fact that imported bioenergy in many cases would be considerably cheaper than what can be domestically produced and support to import could therefore meet political and other resistance from domestic producers.

Another reason may be that it is assumed that support in one link of the chain, e.g. feed-in tariffs, would mobilize the market forces in the entire chain. As has been high-lighted in other parts of this chapter, this assumption would not be true due to the specific financing difficulties for multi-faced bioenergy supply chains for international trade.

Another, probably unexpected, effect of the mandates relates to the roles of big utilities for reaching the mandated targets. In many countries, e.g. the UK, the Netherlands, Denmark, Japan and Korea, power utilities are regarded as key actors in that process. That has given them a strong platform for negotiation with governments, etc. regarding support measures. In some countries this negotiations are on-going and consequently no or little activities take place, in other countries the negotiations have led to satisfactory conditions for the utilities, triggering fast and large-scale actions. Import of bioenergy is an obvious key element in that progress. For these corporations, financing can be organized "in-house", thus the uncertainty factors can be absorbed within the corporation.

Generally, bioenergy trade is not recognized as a key element in present mandates and related policy measures. It seem obvious, that the "market forces" alone would not be able to lead to efficient supply chains. A major factor for the market failures relates to perceived uncertainty, which in turn causes financing problem.

Mandates do work to initiate investments if they are

- long term to allow for sufficient time for the needed investment horizon. The mandates should be designed in order to fit with the policies of the financing sector. For equity financing in long-term projects, the pay-back time is typically 5–10 years; for loans normally somewhat longer. If mandates have shorter horizons, the financing sector would regard that as an uncertainty factor leading to higher rates and shorter amortization times, etc., which in turn will cause i.a. cash flow problems during the most sensitive first years for the investment projects.
- reliable and grandfathered (don't change the rules during the play) (see above) Mandates should be consistent and comprehensive and include visions and explanations on why the mandate is introduced. The development of mandate should be carried out in parallel with the development of measures to gain acceptance from the general public for the need to transform the energy system.
- financially affordable (make sure financing can made available as needed either via flat increase or direct financial incentives and support). Mandates that are based on financial support directly from the state budgets have proven to be shaky in periods of financial crises. In that perspective, other solutions e.g. carbon taxes or quota would be preferred.
- technically achievable (no mandates for products or technologies that are NOT commercially available). It is important to distinguish between support to R&D projects and support to implementation of proven solutions in this case to investments in projects in the supply chain for bioenergy trade. R&D projects may need seed money, and step-wise increases related to progress while investments in the supply chain for proven technologies require support policies like those described above.
- enforceable (have rules to penalize). Mandates that are not enforced via a penalty are no mandates. The penalty has to be well above any potential economic gain that could be achieved in not meeting the mandate.

9.5.3 Possible Ways Forward

In order to enhance the investment climate along sustainable international biomass supply chains a number of needed and/or enhancing prerequisite can be listed that both industry and policy should focus on:

- ***Assure sustainability of feedstock production through certification***. While formal certification might be required in early phases of an emerging bioenergy trade development to motivate political support measures and to safe-guard against abuse, it is likely that the bioenergy trade will develop more transparency and that other measures could replace formal certification, e.g. verification and monitoring carried out by the trading parties and other stakeholders. An established trade would mean that parties in the links have a genuine interest in smooth performance and in avoiding criticism from third parties.

- ***Balance supply and demand through focus on integrated supply chains to develop the market***. Such an approach will inherently guarantee security of supply for each investment and reduce an important risk factor by doing risk so while at the same time reduce financing costs.
- ***Foster commoditization of specification driven bioenergy carriers for international trade through international cooperation on standardization, harmonization.*** This approach would quickly increase the liquidity of the biomass feedstock and bio fuel market and allow for better risk mitigation and sourcing flexibility as well as for more investment into direct feedstock production.
- ***Focus in international bioenergy trade on few preferential bioenergy products that have the potential to be produced, shipped and used in large quantities worldwide***. That focus would allow for an even faster growth of certain supply markets, reduce insecurity within a nascent industry and mobilizes even more financial resources. Economies of scale are significant in the bioenergy value chain and reduced overall costs could eventually lead to independence from any political support schemes in certain energy sectors – as is nowadays happening already for domestic heating applications.
- ***Provide a stable, favourable and reliable economic environment for trade to grow***. The conditions for bioenergy trade have been very different with regard to market incentives, types of fuels, political policies, roles of key actors, etc. Also factors like economic trends and weather have fluctuated. To a large extent it has been the "buyers market's", meaning that much of the risks and uncertainty have been place on early links in the supply chain. The general policy targets set up by UN bodies, EU, etc. provide a structure the development of balanced and harmonized future conditions.
- ***Avoid overcompensation by direct subsidies and stay away as much as possible from the use of food and fodder products to avoid market distortions and negative political repercussions that might result in an abrupt change of policy and mandates***. The long turn aim of support measures, if needed, should be harmonization between all renewable energy systems. The WTO policy should prevail, i.e. not allowing direct subsidies for products subject to international competition. The present controversy about the "food versus fuel" would be reduced by demonstration of positive effects of biomass production for energy, e.g. by application of agro-forestry, utilization of agricultural waste, transferring marginal agricultural land into energy plantations, etc. In particular, the socio-economic effects should be in focus.
- ***Provide better support for the market introduction of new bioenergy trade related technologies through industry consortia, a culture of venture capital and performance based grants***. The application of efficient technologies and

methodologies is a key factor for compatibility of new or expanded projects. However, availability of those techniques and methods is not globally at hand, especially with regard to developing countries. Many, if not most companies in the bioenergy technology industry are small with limited resources for expansion into new markets. This problem is recognized, e.g. in the documents from Climate Change Summits, suggesting that specific fund resources should be earmarked for making technology available for developing countries.
- ***Finally provide a stable regulatory environment that incentives government, industry and private investors to invest along international biomass supply chains***. In a few cases, examples that development has taken place, and others may follow. However, on the policy level, several issues have to be resolved, e.g. support to domestic supply chains versus import chains, application of quota or incentive measures ("stick or carrot"), and focusing on advanced or proven technologies.

9.6 Synthesis

Rising fossil fuels prices, ever more efficient production, transportation and conversion of biomass based fuels and the growing political will to reduce CO_2 emissions will lead to a continuous rise of the use of biomass products also in the energy sector worldwide. In order to balance demand and supply the share of internationally traded biomass fuels will even increase in the future.

This effect offers a wide range of investment opportunities along the whole value chain of international biomass supply. But biomass to energy value chains are far more complex than any other form of renewable energy and pose a whole new set of ecological, technical, logistical and commercial challenges. So financing and managing the growth of this new industry is not easy and will take more time in relative terms but reach higher levels in absolute terms in comparison with other renewable energy sectors.

The growth of sustainable energy biomass production, use and international biomass trade is a very positive trend that needs to be supported and steered by far sighted policy making sure that policy measures give enough support to biomass based fuels to assure their long term competitiveness with fossil based alternatives as well as their overall sustainability.

Overall policy measures to increase the use of biomass for energy certainly did lead to an increase in bioenergy use so far and as long as policy support is needed to support the development of this energy sector the more these policies have to gain the trust of the investment community in order to provide for the needed long term finance.

Biomass is the only primary renewable energy source that can be stored and transported over long distances. Its overall worldwide sustainable potential is large and it has the potential to serve not only the energy market. In light of these fundamentals investments in biomass supply chains offer a very positive long term perspective.

Appendix

Selected general internet links

- http://www.risiinfo.com/
- http://www.hawkinswright.com/
- http://www.crossborderbioenergy.eu/
- http://about.bnef.com/

Selected funding agencies in different countries:

USA:

- USDA Business & Industry Guaranteed Loans (B&I Loans) http://www.rurdev.usda.gov/rbs/busp/b&i_gar.htm
- USDA Renewable Energy for America Program (REAP) Guaranteed Loans http://www.rurdev.usda.gov/rbs/busp/9006loan.htm
- Biorefinery Assistance Program http://www.rurdev.usda.gov/rbs/busp/baplg9003.htm
- DOE Innovative Technology Loan Guarantee Program http://www.lgprogram.energy.gov/features.html

Africa:

- African Development Bank, www.afdb.org

Asia:

- Asian Development Bank, www.adb.org

EU:

- EBRD, www.ebrd.com
- EIB, www.eib.com

Germany:

- KfW, www.kfw.de
- GIZ, www.giz.de

References

AEBIOM. (2012). *2012 European bioenergy outlook*. Available at: http://www.aebiom.org/?p=6400#more-6400. Accessed Feb 2013.

Bradley, D., Hektor, B., & Schouwenberg, P. P. (2010). *World bio-trade equity fund study*. Available at: http://bioenergytrade.org/downloads/biotradeequityfundfinalreport.pdf. Accessed Feb 2013.

Cocchi, M. et al. (2011). *Global wood pellet industry and trade study*. Available at: http://bioenergytrade.org/downloads/t40-global-wood-pellet-market-study_final.pdf. Accessed Feb 2013.

Deutmeyer, M. et al. (2012). *Possible effect of torrefaction on biomass trade*. Available at: http://bioenergytrade.org/downloads/t40-torrefaction-2012.pdf. Accessed Feb 2013.
IEA. (2007). *Renewables in global energy supply*. Available at: http://www.iea.org/publications/freepublications/publication/renewable_factsheet.pdf. Accessed Feb 2013.
Junginger, M., van Dam, J., Zarrilli, S., Mohamed, F. A., Marchald, D., & Faaij, A. (2010). Opportunities and barriers for international bioenergy trade. *Energy Policy, 39*(4), 2028–2042.

Chapter 10
Synthesis and Recommendations

Martin Junginger, Chun Sheng Goh, and André Faaij

Abstract In this chapter, the main insights and lessons of the book are synthesized, providing an overview of trade flows and key findings on amongst others the importance of biomass trade in relation to policies, logistics, sustainability. Based on that, recommendations are formulated for policy makers, NGO's, industry and academia how to further develop international bioenergy trade. The chapter is closed by a vision on future bio-based economy, international markets and trade. We show that while bioenergy trade is not a goal on its own, and that substantial challenges for the future development of global and international bioenergy trade may be expected in the coming decades, trade is a crucial prerequisite to balance demand and supply on an international scale and in a sustainable manner which will lead to the development of bioenergy as a carbon-neutral alternative to fossil fuels. At the same time, trade can mobilize rural areas around the world into becoming key energy producers and exporters which in turn can contribute to poverty alleviation and further development and modernization of the agricultural sector. However, such developments will need governance frameworks and best industry examples to ensure sustainable production, trade and use – and this book provides many state-of-the-art insights in what can be learned from current market experiences, certification efforts, logistics and preconditions with respect to policy and investment.

M. Junginger (✉) • C.S. Goh • A. Faaij
Copernicus Institute, Utrecht University, Utrecht, The Netherlands
e-mail: h.m.junginger@uu.nl; c.s.goh@uu.nl; a.p.c.faaij@uu.nl

M. Junginger et al. (eds.), *International Bioenergy Trade: History, status & outlook on securing sustainable bioenergy supply, demand and markets*, Lecture Notes in Energy 17,
DOI 10.1007/978-94-007-6982-3_10,

10.1 Main Insights and Lessons from This Book

10.1.1 Bioenergy Trade in a Nutshell

While the overall use of modern bioenergy has increased gradually over the past decades in many countries of the world, international bioenergy trade is a phenomenon that has grown from virtually nothing to substantial volumes in little more than a decade (2000-current), with internationally traded volumes typically increasing by a factor of 10 or more over this period. Main biomass fuels traded over long distances are liquid transport fuels (ethanol and biodiesel) and refined solid biomass (wood pellets and to a lesser extent wood chips). At the moment, liquid and solid biofuels markets are significantly different. The liquid biofuels markets are reasonably developed markets and are closely related to agriculture commodities. Solid biofuels basically originated from the forestry and wood processing sector, and are mainly used in renewable electricity and heat production. The markets are less complex and trade dynamics are more straightforward. By 2011, close to 2,500 PJ of liquid biofuels were produced globally; over two-third of which were fuel ethanol and the remaining biodiesel. About 300 PJ of liquid biofuels were traded internationally in 2011, of which roughly two-thirds biodiesel. The feedstock base is exclusively regionally specific oil, sugar, or starch crops. Global trade in biodiesel has been and will in the foreseeable future be primarily driven towards the European Union, where renewable energy policies stimulate the consumption of sustainable transport fuels – although the EU biofuels market growth is slowing down. Fuel ethanol is largely produced and consumed in the Americas, with the USA and Brazil dominating global production, trade and deployment. In comparison, solid biomass trade for energy has also reached over 300 PJ by 2011. The majority of this volume comprises of wood pellets and wood chips aimed for consumption in the European Union (EU). Wood pellets are the largest single commodity stream and have seen a rapid production growth and trade internationalization. This is primarily due to past and expected future EU demand developments in the industrial segment. Belgium, the Netherlands, the United Kingdom, and Denmark in particular are bound to increase consumption, and will remain net pellet importers. To a lesser extent, solid biomass is also traded in unrefined (such as palm kernel shells). Also, indirect trade of biomass is significant: large amounts of biomass is traded primarily for material purposes (e.g. sawn wood, pulp, animal fodder), but after further processing and final use in the destination country, large amounts of these streams still end up as fuel in the wood and paper industry or in municipal solid waste, which is increasingly combusted with electricity and/or heat production.

10.1.2 Policy as Main Driver

It is obvious from many chapters in this book that policies supporting the increase demand in biomass as part of renewable energy policies (both liquid biofuels and solid biomass) have been the single-most important driver behind this increase in trade. Other potential drivers (increased fossil fuel prices, prices for CO_2, security of supply/geopolitical concerns, policies to stimulate biomass production/export) have all been (far) less important to date. While these are expected to increase in importance in the more distant future, up till 2020, the renewable energy targets in the US and the EU are likely to remain the largest single driver for international bioenergy trade until 2020. Ambitious targets in almost all OECD countries are likely to lead to further increasing imports of biomass in the near future. While in the previous decade, increasing oil- and coal prices fostered the hope that renewables would become cost-competitive o a direct comparison (i.e. without taking externalities into account), this has become less likely with the effects of (cheap) shale gas and oil also indirectly lowering the prices of coal. Maintaining renewable energy targets (and effective support policies to reach them) is thus of undiminished importance.

10.1.3 The Importance of Sustainability

The main rationale for bioenergy deployment is to enable society to transform to more sustainable fuel and energy production systems. Thus, sustainability safeguards are needed. These can be either through binding regulations and/or voluntary systems, both for domestic and imported biomass. There is currently a high number of initiatives and a proliferation of schemes. Markets would gain from more harmonization and cross-compliance, but also need a good balance between complexity and accessibility. If too many or too complex indicators are implemented, the certification process becomes too demanding, costly and difficult to manage and thus not attractive for users. Too little detail will lead to different interpretation of the principles and may increase the risk of 'green-washing'. A common language is needed as 'sustainability' of biomass involves different policy arenas and legal settings. Standardization has proven to be very important (in other sectors) to create transparent markets and thereby facilitate rational production and trade. Design of sustainability assurance systems (both through binding regulations and voluntary certification) should take into account how markets work, in relation to different biomass applications (avoiding discrimination among end-uses and users). It should also take into account the way investment decisions are taken, administrative requirements for smallholders, and the position of developing countries.

Sustainability requirements are evolving and discussions on topics like iLUC for biofuels or carbon accounting for solid biomass are creating high uncertainties for companies, which in the future may need to comply with sustainability requirements that are unknown today.. The developments in biofuels markets (such as the recent change in EU policy limiting first generation biofuels to a contribution of 5 %) show clearly that uncertainty and ongoing changes in policies and regulations cause markets to stagnate. Similarly to the lengthy ongoing debates over iLUC risks for biofuels, the carbon accounting debate for forest biomass is appearing to head in the same direction. It should be kept in mind that stakeholders are making investment decisions now which establish long-term contracts, whereas governments may evaluate their policy year by year. Therefore, policy pathways should be clear and predictable, and future revisions of sustainability requirements should be open and transparent.

10.1.4 The Role of Efficient Logistics and Pretreatment Technologies

Achieving objectives for increased use of renewable bioenergy resources will require that biomass has the characteristics needed to be bought and sold outside of its production areas, or that biomass is "tradable." Tradability is influenced by the reliability of product supply, the existence of a market demand, the opportunity for profitable transactions, the physical transferability of the product, and the guarantee of product quality. These influences are not exclusive of one another, and the more that must be done to a product to improve its tradability, the greater the cost constraint pressures become. The fundamental challenge facing an expanding bioenergy industry is that feedstock cost is presently too high for demand. A key component of reducing biomass cost is to change the format to something high-density, flowable, stable, consistent, and high quality. Making the biomass format compatible with existing high-capacity transportation and handling infrastructure will reduce the need for new infrastructure. Producing biomass with these characteristics and a cost conducive to energy production requires the development of new technologies, or improvements to existing ones.

One of these new technologies that can make biomass 'tradeable' may be torrefaction. It is a thermochemical treatment that drives off moisture and volatiles, and produces an energy-dense, homogenous, stable, and easily transported feedstock product. The product lacks bulk density, which can be improved using densification technologies such as pelletization to increasing dry matter bulk density and address feedstock mobility and durability issues that are coupled to long-distance transport and logistics. There are many advantages to densification, including improved handling and conveyance efficiencies, controlled particle size distribution for improved uniformity, improved stability, quality improvements, and improved performance in conversion systems. Currently, however, these technologies are often energy intensive and too costly.

10.1.5 Tradeable Feedstocks and the Biorefinery

As demand for bioenergy increases, the amount of feedstock resources required to support production will be significant. To meet feedstock needs, a typical biorefinery may receive a variety of feedstocks, e.g. switchgrass, corn stover, miscanthus and/or eucalyptus, depending on location and availability. These feedstocks vary widely in composition and recalcitrance, and would require biorefineries to optimize (and possibly re-engineer) their processes for each different type of biomass, thus increasing costs. Complicating this further is that feedstock diversity varies markedly from region to region, and each feedstock within a region varies from year to year based on weather conditions, handling, storage, and crop variety. This will result in different types of biorefineries needed in every region which will further increase costs for construction and operation since there will be no "standard" biorefinery. Least-cost formulation, in conjunction with mechanical preprocessing and preconversion technologies, offers a promising solution to these issues by combining feedstocks to achieve desired feedstock specifications, reduce undesirable properties, and simplify downstream processing. This approach leverages technology advances to address transport and logistics challenges and convert raw biomass into feedstocks that are easily traded outside their production areas.

10.1.6 Barriers

Barriers restricting the development of bioenergy trade include technical and tariff barriers currently mainly affect liquid biofuels. Import tariffs and anti-dumping measures have been the topic of dispute between the main producing and consuming regions of ethanol and biodiesel for the last decade, and also technical standards for biodiesel have been criticized, as they may put biodiesel made from soy and palm kernel oil at an disadvantage. For solid biomass, phytosanitary measures are one of the most important barriers preventing the trade of softwood wood chips for energy. Also health and safety issues related to transporting and storing solid biomass still need further attention. Finally, also the lack of good biomass (liquids and solids) production, trade & use statistics as a limiting factor to trade is mentioned in several chapters.

10.1.7 Future Demand

While international biomass trade is likely to continue to grow, it is questionable whether it will be able to maintain its exponential grow for the next decades. This would however be necessary to reach the amounts required in several model scenarios, which display large regional demand vs. supply gaps – a fact which is currently likely underestimated by policy makers and modelers etc. alike. Without significant bioenergy trade between world regions, no (or only much more limited) significant growth of bioenergy is achievable. Hence, either major challenges regarding social, ecological, technical, logistical, economic aspects of international bioenergy trade will have to be solved, or the objectives of significant higher bioenergy use have to be reduced.

10.1.8 Economic Opportunities

Setting up long-distance supply chains may then also offer a broad array of interesting investment opportunities that could offer stable, long term and high returns. But under the current investment climate in many parts of the world, investment in bioenergy supply chains is perceived by the finance sector to be risky and uncertain and therefore many projects do not materialize. At present, insufficient financing is an obvious obstacle, and so far mainly large energy companies are the largest single group of investors into bioenergy and international bioenergy supply chains. Institutional and private investors will have to step up and increase investments in bioenergy trade – possibly through dedicated bioenergy trade equity funds, and policy makers would have to create the boundary conditions (through favorable policy and investments in infrastructure). Other bioenergy market parties, e.g. the small and medium scale heat markets, small fuel producers, independent agents and distributors will also have to become more actively involved in international bioenergy trade.

10.2 The Way Forward – Actions Required from Stakeholders

Further expanding bioenergy trade is thus an extraordinary challenge, which will require action from all stakeholders involved. Below, we discuss in more details what will be needed from each stakeholder group to make this happen.

10.2.1 Policy Makers

Given the continued dependence of renewable energy sources on policy support, first and foremost, there is a strong need to set clear and binding long-term targets for renewable energy and GHG emission reduction targets, especially beyond 2020, preferably on an international level. Linked to this, also the framework of economic support policies should be permanent and internationally harmonized. International mandates for the use of renewable energy in general and bioenergy specifically focusing primarily on the transportation, heat and power sectors have proven to be effective, as have direct subsidies in individual investments along the biomass to energy value chain. National stop-and-go policies have proven to disrupt the market significantly.

In parallel, also an internationally widely accepted framework on the sustainable production of biomass for energy, material use and other purposes should be established. While scientific insights on the sustainable production of biomass progress over time, and sustainability criteria may be sharpened as the industry learns, again, such changes should be introduced gradually, with sufficient time for all stakeholders in the supply chain to adapt.

As sustainability of biomass and bioenergy is politically important in the context of carbon mitigation policies, transparent and consistent monitoring systems should be established to minimize the risk of confusion, improve the reliability of trade information, facilitate the discussions between policy makers, NGOS and industry, and support sustainable international bioenergy trade.

Next to these general recommendations to promote the sustainable production and use of bioenergy, policy makers must also realize the logistic challenges that go hand in hand with developing large-scale bioenergy trade. Biomass trade will have to reach the dimensions similar to those of coal or oil today, which presents a tremendous challenge, as biomass is typically produced decentralized, and thus needs to be collected form wide and should enable and stimulate investments in infrastructure, especially in exporting countries, e.g. by creation of suitable import and export legislation for bioenergy products. In addition, countries that currently still have tariff barriers in place may consider to remove them. If done globally, this would create a more level playing field for producers, and could ultimately also lead to lower cost of policy support measures.

Linked to the logistic challenges, support for the development of new technologies for efficient collection, pretreatment and transport of bioenergy is indispensable to further lower the costs of logistics on the longer term.

10.2.2 NGO's

Many NGO's have in recent years increasingly opposed bioenergy in general; first focusing on first generation liquid biofuels, and more recently also on the use of forest biomass. Underlying concerns are impacts on biodiversity, food security and timing before GHG savings are obtained. While these concerns are in principle all valid, and safeguards have to be established to prevent the use of unsustainably produced biomass, NGO's have a tendency to (unknowingly or deliberately) sketch worst case scenarios, which have often little or no relevance to actual ongoing industry practices. One prominent example is the carbon debt debate, where opponents often assume that entire old-growth forests are integrally cut down and used for energy and heat production, whereas in reality, typically only residues and thinnings from plantations and sustainably managed forests are used, and these – at least currently – only contribute an very minor share of the modern bioenergy use in OECD countries. Similarly, NGO's have often objected against the long distance trade of bioenergy, claiming that energy losses would be prohibitive, while scientific studies have shown repeatedly, that long-distance trade – if organized right, does not cause prohibitive energy losses, i.e. 10–20 % of the energy content, which is no more than e.g. electricity losses from offshore wind farms transported long-distance to demand centers, or solar electricity transported from the Sahara to Southern Europe. While the importance of NGO's to point out unsustainable practices and black sheep is not disputed, their aims and actions in recent years seem to have gone too far. We therefore encourage NGO's to be more pro-active: to accept bioenergy in general as part of the solution rather than of the problem, and focus on how sustainable production and trade (including ecological and social benefits) can be established. Such pro-active roles can for example be achieved by participating in roundtables on the development of certification schemes.

10.2.3 Academia

Task 40 has had a large number of academic institutes amongst its member since the start, and for good reasons. With the development of bioenergy trade, all kinds of questions have risen (and continue to rise), as diverse as the development of effective measures against health and safety hazards, setting up reliable bioenergy trade statistics, delivering integrated assessments of biomass producing & sourcing regions (including economic, social and environmental aspects), developing advanced biomass pretreatment methods et cetera. Nevertheless, science has in some ways also not been able to deliver: for example, regarding both the iLUC and the carbon accounting of forest biomass, different scientific views exist, and there is little consensus on the correct calculations methods and choice of reference systems. There is a clear need for scientists to both highlight the uncertainties in the current studies and at the same time improve these tools to provide policy makers and other stakeholders high-quality information for decision making. As a general recommendation, also scientists should communicate (better) with other stakeholders, especially industry, to signal early warning lights, but also to learn what current industry practices are and what questions from market actors need to be answered.

10.2.4 Industry

Last but not least, the industry itself is called upon as the most important party to further develop bioenergy trade. As already argued in Chap. 9, industry will need to assure sustainability of feedstock production through certification. While formal certification might be required in early phases of an emerging bioenergy trade development to motivate political support measures and to safe-guard against abuse, it is likely that the bioenergy trade will develop more transparency and that other measures could replace formal certification, e.g. verification and monitoring carried out by the trading parties and other stakeholders. An established trade would mean that parties in the links have a genuine interest in smooth performance and in avoiding criticism from third parties.

The industry will also need to balance supply and demand more carefully through focus on integrated supply chains to develop the market. Such an approach will inherently guarantee security of supply for each investment and reduce an important risk factor by doing risk so while at the same time reduce financing costs.

Another key aspect is to foster commoditization of specification driven bioenergy carriers for international trade through international cooperation on standardization and harmonization. This approach would quickly increase the liquidity of the biomass feedstock and biofuel market and allow for better risk mitigation and sourcing flexibility as well as for more investment into direct feedstock production.

Finally, the focus will need to be put on a few preferential bioenergy products that have the potential to be produced, shipped and used in large quantities worldwide. This would allow for an even faster growth of certain supply markets, reduce insecurity within a nascent industry and mobilizes even more financial resources. Economies of scale are significant in the bioenergy value chain and reduced overall costs could eventually lead to independence from any political support schemes in certain energy sectors – as is nowadays happening already for domestic heating applications.

10.3 Closing Remarks; a Vision on Future Bio-based Economy, International Markets and Trade

In the introduction of this book, a mixed picture was sketched with respect to future biomass deployment: In state-of-the-art energy and GHG mitigation scenario's, biomass plays an essential role in meeting ambitious GHG mitigation targets, in particular for transport fuels (including growing demand from aviation and shipping) and feedstocks for chemical industry. Biomass shares in total energy supply are hoped to achieve 20–30 % looking at long term strategies and scenarios of countries as well as on a global scale. Recent scenario's stress the importance of combining large scale bioenergy use with carbon capture and storage, because this is one of the few options available to achieve net negative emissions and keep a global 2 °C temperature change target in sight during this century; a target widely seen as essential to limit the damage of climate change (GEA 2012).

Opportunities achieving synergies with rural development and ecosystem services are key benefits that could be realized in conjunction with such large scale bioenergy deployment. But achieving such an optimal outcome, comes with a large number of preconditions, most notably with respect to governance of land use & natural resources and modernization of agriculture. Furthermore, in many areas improved technologies and more experience is needed to improve environmental performance and economics, This is true for advanced technologies (fuels, biomaterials, power, carbon management such as the combination of bioenergy and carbon capture and storage), cropping systems in many different settings, further development and optimization of infrastructure, logistics, functional markets and chain of custody control. Last but not least, effective business models & investments are required.

It is by no means certain that such a mix of preconditions will be secured, despite the strong drivers to do so. Given the expectations for a high bioenergy demand on a global scale, the pressure on available biomass resources will increase. Without the further development and mobilisation of biomass resources (e.g. through energy crops and better use of agro-forestry residues) and a well-functioning biomass market to assure a reliable and lasting supply, those ambitions will not be met. Deployment of biomass may therefore remain fairly low at a 100 EJ level in 2050 as argued by the IPCC Special Report on Renewable Energy report. This is

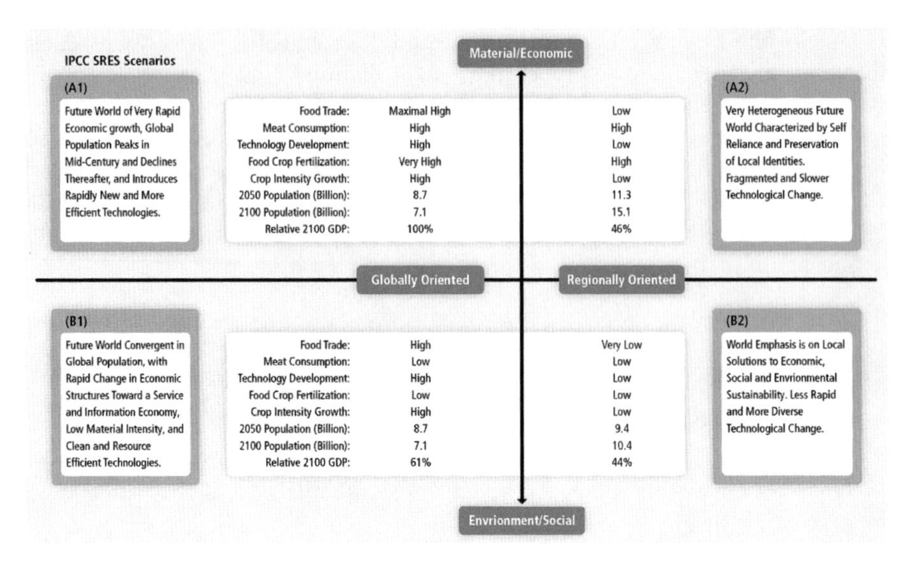

Fig. 10.1 Storylines for the key scenario variables of the IPCC Special Report on Emissions Scenarios (SRES)

illustrated by Figs. 10.1 and 10.2 (Chum et al. 2011) that sketches the "scenario space" for different futures (the so called SRES scenario's). Figure 10.1 gives the general storylines. Figure 10.2 sketches a number of main possible characteristics with respect to bioenergy deployment, related impacts and preconditions (or lack of preconditions) for the different storylines. Of course, these are scenario's. In reality, the development pathways that could be followed could be profoundly different going from region to region across the globe and over time. Therefore, predicting future bioenergy markets and trade remains uncertain.

Model scenarios with an ambitious increase of bioenergy demand imply a huge increase in bioenergy trade, an increase by a factor of 70 between 2010 and 2030 for liquid biofuels, and by a factor of 80 for solid biomass between world regions. Such an increase would result in quantities of internationally traded biomass commodities which would be higher than the current total global bioenergy demand (i.e. larger than 50 EJ). Considering the currently very small share of internationally traded bioenergy, this would result in huge challenges and tremendous changes in terms of production, pretreatment of biomass and development of logistic chains.

At the same time, the level of international bioenergy trade shown in the model scenarios is necessary to fill the anticipated regional gap between demand and supply. Without significant bioenergy trade between world regions, a much less pronounced growth of bioenergy is achievable.

The insight into future scenarios and perspectives of bioenergy trade makes clear that substantial challenges for the future development of global and international bioenergy trade may be expected in the coming decades if a low carbon energy system is to be implemented globally. Some of these, such as the development of

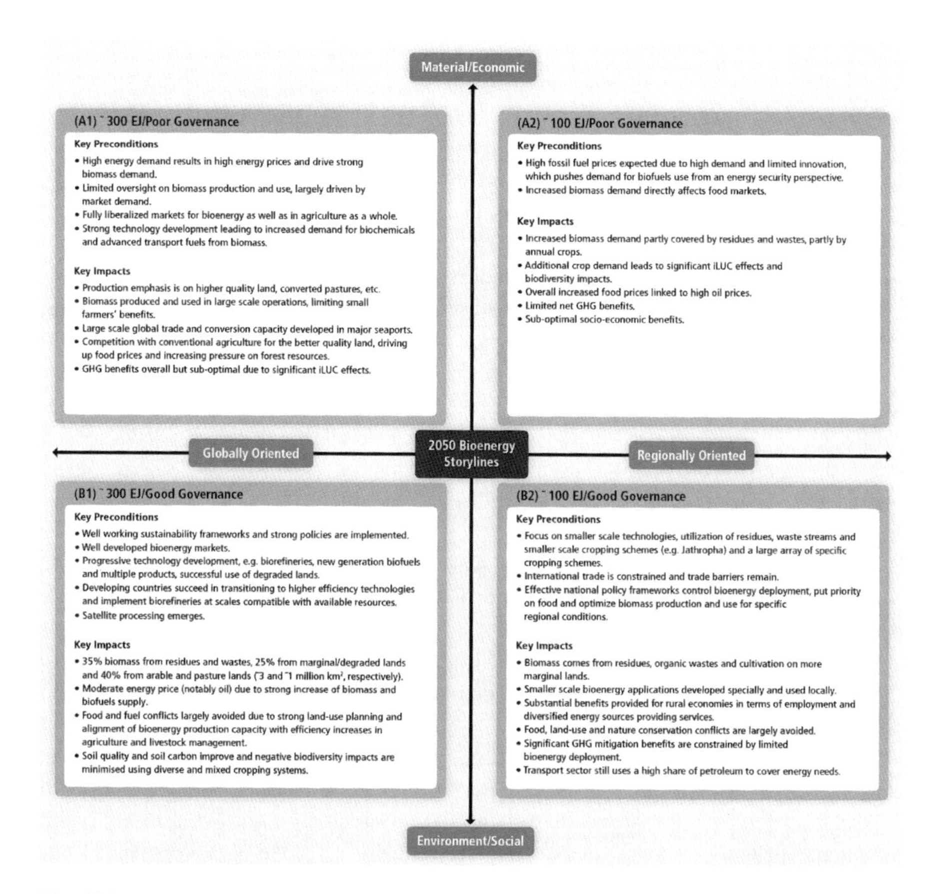

Fig. 10.2 Possible futures for 2050 bioenergy deployment: Four illustrative contrasting sketches describing key preconditions and impacts following world conditions typical of the IPCC SRES storylines (IPCC 2000)

logistics, the required investments to realize production and trade, and the need to govern sustainable production of bioenergy are addressed in this book. Others are still open for further research, e.g. the implications of bioenergy trade for specific regions and for different biomass commodities in terms of social, ecological and economic impacts or the effect of fluctuating exchange rates, regional development of economic and policy side conditions.

Biomass trade is not a goal of its own. But balancing demand and supply on an international scale and in a sustainable manner will lead to further CO_2 emission reductions and offer new chances for rural regions, degraded areas worldwide and in particular in developing countries. In that sense, the developing international bioenergy markets also represent an extraordinary opportunity: If indeed the global bioenergy market is to develop to a size of 300 EJ per year over this century (compared to 500 EJ current total global energy use), the value of that market at US$ 6/GJ (considering pre-treated biomass such as pellets) would amount to some US$ 1.8

trillion per year. Logically, not all biomass will be traded on international markets, but such an indicative estimate makes clear what the economic importance of this market can become for rural areas worldwide, as are the employment implications.

It is exactly there that the possibilities and potentials for modern bioenergy production, including export, are largest and at the same time the need for development of rural areas is the highest. These crucial issues, global bioenergy markets and rural development, merge in a formidable way.

Given the scale of the market, bioenergy trade could provide one of the most important sustainable development pathways for decades to come: developing bioenergy as the key sustainable and carbon-neutral alternative to fossil fuels and at the same time mobilizing rural areas around the world into becoming key energy producers and exporters could contribute to poverty alleviation and further development and modernization of the agricultural sector. The latter is extremely important for sustainable development of agriculture and food security at large, while lowering the footprint for food production and allowing biomass production without unsustainable land use change.

Although biomass production may well provide a crucial strategy to enhance sustainable land-use management, negative developments should be avoided, e.g., by clear standards and best-practice guidelines for (the design of) biomass production systems and their integration in agricultural areas.

Gaining further experience with certification, developing and implementing the necessary governance frameworks on land use, removing trade barriers and showing best-practice operations through export-oriented production schemes in a diversity of developing countries and different rural areas are crucial in the short term. This book has provided many state-of-the-art insights in what can learned from current market experiences, certification efforts, logistics and preconditions with respect to policy and investment. Good examples, successful business models and effective frameworks can guide market forces in a sustainable direction. When successful, the Green OPEC (or BIO-PEC) may become a reality in several decades from now!

References

Chum, H., Faaij, A., & Moreira, J., (CLA's), Berndes, G., Dhamija, P., Dong, H., Gabrielle, B. X., Goss Eng, A. M., Lucht, W., Mapako, M., Masera Cerutti, O., McIntyre, T. C., Minowa, T., Pingoud, K., Bain, R., Chiang, R., Dawe, D., Heath, G., Junginger, M., Patel, M., Yang, J. C., & Warner, E. (2011). Chapter 2, Bioenergy. In O. Edenhofer, R. P. Madruga, & Y. Sokona et al. (Eds.), *The IPCC special report of the* Intergovernmental Panel on Climate Change*: Renewable energy sources and climate change mitigation*. New York: Cambridge University Press. Available at: http://srren.ipcc-wg3.de/report/IPCC_SRREN_Ch02.pdf

GEA. (2012). *Global Energy Assessment – Toward a sustainable future*. Cambridge/New York/Laxenburg: Cambridge University Press/The International Institute for Applied Systems Analysis.

IPCC. (2000). *Summary for policymakers. Emissions scenarios*. A special report of IPCC Working Group III. Available at: http://www.ipcc.ch/pdf/special-reports/spm/sres-en.pdf

Index

M. Junginger et al. (eds.), *International Bioenergy Trade: History, status & outlook on securing sustainable bioenergy supply, demand and markets*, Lecture Notes in Energy 17, DOI 10.1007/978-94-007-6982-3, © Springer Science+Business Media Dordrecht 2014

Printed by Printforce, the Netherlands